THE SURPRISING WORLD OF ZERO

The emancipation
of the First Cardinal

CAROLINA PARIS

Special Thanks to Sara Contini

ISBN-10: 1534838686
ISBN-13: 978-1534838680

TO GIACOMO LEOPARDI

Does time have a meaning?
There is "before" and there is "after", there is "to be there"
and "not to be there", but how much time does one need to
understand a moment? Concentration fails me, as do the
moments I do not understand. I am looking for a way, I watch
time and space merge, they get lost without a solution, without
a concept to express them. Help, I feel like begging for some
help. Help! To find a way back. I am looking for the tender
that navigates the waves, light and sure, to drive this wrecked
ship to a safe harbour. This is the time for peace and serenity.
It is time. This is time, whatever that means. But what if it
was not? What if there was no "before", no "after"? If
existence was not a straight line, but a circle? Your "after"
might be someone else's "before". If you are in a circular
system-existence, the only relationship, the only confrontation
may have is with the centre, which is not above nor under, but
on the same level. What if the circle was part of a sphere? Once
again, you may compare yourself to the centre, if you wish. One
sphere, thousands, infinite circles. Outside each circle, but
inside the sphere — which means inside each possible circle —
there is the centre, transcendent and immanent. A single centre
for an infinite variety of system-existences. You are a point that
travels through part of a circle. Not as important as you might
think, but essential nonetheless to the existence of the circle and
of the sphere. All so that the centre has a meaning. Unless you
are the centre, be satisfied with the small tract of the circle that
you are allowed to travel and enjoy the importance of that tract
for the sphere and the centre.

Index

1 THE JOURNEY BEGINS 7

2 EXPOSING THE ZERO...13

3 ZERO AS NOTHING, NOTHING AS EMPTINESS 15

4 NOTHING AS NULL...21

5 ZERO AS OPPOSED TO ONE 23

6 NOTHING AS LACK OF SOMETHING 27

7 FROM PARMENIDES TO PLATO 31

8 ARISTOTELES' VISION 37

9 NOTHING AND NIHILISM..41

10 THE TRAGIC NOTHINGNESS OF EXISTENTIALISTS...................45

11 FROM PHYSICS TO METAPHYSICS 49

12 NOTHINGNESS IN RELIGION...55

13 THE WORLD FROM NOTHING 63

14 A MORE CONCRETE NOTHINGNESS 67

15 THE MATHEMATICAL STANDPOINT....................................69

16 COUNTING SYSTEMS..77

17 A WORLD WITHOUT THE ZERO? 83

18 THE DECIMAL POSITIONAL   SYSTEM 89

19 ZERO BECOMES A NUMBER 93

20 OPERATIONS WITH ZERO 97

21 CALCULUS...107

22 MISTREATED ZERO 111

23 RABBINIC HEBRAISM AND KABBALAH 119

24 THE FULNESS OF ZERO 121

25 THE PRIMORDIAL EGG 127

26 ZERO IN MANTIC ARTS...135

27 ZERO EMBRACES TRASCENDENCE..141

28 THE EMANCIPATION OF ZERO 151

29 AFTERWORD 161

30 BIBLIOGRAPHY AND WEB LINKS 169

31 ABOUT THE AUTHOR 174

1 THE JOURNEY BEGINS

Counting is the essence of life. For many people. Luckily, not for everybody.

It is necessary to be One. Naturally, one that counts. You are taught to aspire to that status since childhood.

Therefore, you count: one, two, three... You count as much as you can, in order to be always... someone that counts.

It goes without saying, the more you count (money, acquaintances that "count"), the more you are a Number 1, and the Zeros will be willing to take your side. Here is a One that feels like he is Everything. An Everything around which the world revolves, or at least a part - the larger part - of it. The part that puts the One in the centre.

It also goes without saying that according to this logic the person who can count up to 1000 is worthier than one who can hardly get to 100. Not to mention those who reach one million, one billion, or more.

People count and they identify the richest person in the world. They count to establish who is the most powerful, counting all the people that count...

They count according to a cardinality.

In set theory, the term "cardinality" (or "numerosity", "magnitude") indicates the number of elements contained in a finite set.

Georg Ferdinand Ludwig Philipp Cantor, from now on just "Cantor", was born in Saint Petersburg in 1845, but he lived mainly in Germany until his death in 1918. He was a great mathematician and he is considered to be the father of modern set theory.

Cantor expanded the axiomatic set theory, which was developed at the same time, to include the concepts of transfinite numbers, cardinal numbers and ordinal numbers. This is fundamental to our research. Cantor has taught us the cardinality of numbers, which means, in a nutshell, that two is "better" (meaning that it has more value) than one (1), and that one hundred, one thousand, one million, is definitely "better" than one (1).

But in this world seemingly made of cardinals (1, 2, 3...), what should we do with Zero?

The Zero is seemingly ignored, or worse, despised. But that is just a huge injustice that comes from ignorance.

The poet Trilussa was born in Rome as Carlo Alberto Camillo Mariano Salustri (1871). He was also a writer and a journalist. He is well known for his works in the Roman dialect.

Here is how we described his perception of the Zero in the poem entitled *Numbers.*

NUMMERI

Conterò poco, è vero:
-diceva l'Uno ar Zero-.
ma tu che vali? Gnente: propio gnente.
Sia ne l'azzione come ner pensiero
rimani un coso voto e inconcrudente.
Io, invece, se me metto a capofila
de cinque zeri tale e quale a te,
lo sai quanto divento? Centomila.
È questione de nummeri. A un dipresso
è quello che succede ar dittatore
che cresce de potenza e de valore
più so' li zeri che je vanno appresso.

NUMBERS

I am worth little it is true
-said One to Zero
But what are you worth? Nothing, absolutely nothing.
Both in action and in thought,
you remain an empty and ineffectual thing.
Meanwhile if I put myself in front
of five zeros like you
Do you know what I become? One hundred thousand.
It's a question of numbers. That is roughly
what happens to a dictator
who grows in power and value
with the more zeros that follow him.

In this poem, written in 1944, the poet considers the Zero to be worthless, unless it is associated to something else

In that case, speaking of numbers, the Zero, while it does not hold value in itself, can give a specific value to the first number in the line.

The poem was written during WWII and Trilussa, always caustic, was hinting at the fascist regime with this metaphor. He used to describe himself as a non-fascist, rather that antifascist, and while he was not at odds with the regime, he nonetheless felt free to express himself through sharp, explicative satire.

Was it bravery? Or did he just not care about possible consequences? Did he not fear exile? Not exactly. There was some sort of "non-belligerence" pact, a kind of mutual tolerance between Trilussa and the Fascist Party, so that the former could afford to take some jabs at the regime without facing any real consequence.

Aside from the political implications, what really matters here is that the poem is a clear example of a propension (definitely not exclusive of Trilussa) to consider the Zero as something *de facto* equal to Nothing.

The Zero, its equivalence to Nothing, and the Pindaric flights that connect, or better would connect, the Zero to some philosophical or religious equations, have been the subject of thousands of texts, in many different historical periods, in many different languages.

This essay is semi-serious indeed, and rather

unpretentious: its only purpose is to debunk, shatter, smash, destroy the concept of the Zero as Nothing and the idea of Zero as an instrumental number, which implies that Zero holds no value for itself. Of course the Zero, in its sheer numerical acceptation, is indeed instrumental, but its essence and theoretical value are much more than just that.

Our purpose here is to finally restore the Zero's reputation, honor, and, last but not least, splendour.

What we are suggesting here is a point of view, merely a point of view. We are definitely not aiming to compete with the Copernican revolution, or match the creativity of Galileo's visionary mind.

To tell the truth – and we are not being sarcastic – even if we were to produce a completely serious essay, we would not take Galileo as an example.

Galileo Galilei, considered the father of modern science, was born in Pisa in 1564. He studied physics, philosophy, mathematics and astronomy.

Undoubtedly, his contribution to the astronomic revolution is invaluable. He strongly advocated the heliocentric theory of Copernicus, regarding the motion of celestial bodies, over the geocentric theory, supported by the Catholic Church.

But, let's face it, he was also a bit of a coward. He was taken to trial for heresy in 1633. The holy tribunal of the "Santo Uffizio" sentenced him to life in prison and, in order to avoid that, Galileo decided to recant everything he stated in his books and to manifest his full support to the Church and its "scientific" understanding of the world and of the laws of the universe.

There is no cowardice here. We just want to express our point of view. That is all we want. Some may find this point of view debatable, but if it is embraced, it can definitely help to attain a happier approach to life. Without any rhetoric or conceit. Why not try?

2 EXPOSING THE ZERO

Here is our Zero: we want to expose its essence and then dress it again with the most noble and precious garments.

It is easy, for our adult mind, to picture infinity (not in a strictly metaphysical sense) as something enormous, immense, and it is equally easy to picture Zero as something so infinitely small that it equates to Nothing.

But the story of the Zero needs to be rewritten and what we want to express here, without restraints, uncertainty or cowardice of any sort, is an uncontrollable, certainly manifest and heart-felt, apology of the Zero.

We are going to define and suggest some milestones to finally, and deservedly, achieve the emancipation of the Zero!

The first thing we need to assess for our analysis is the fact that over the centuries and millennia many different schools of thought have expressed their views on the Zero. We might say that we had, and still have, different approaches to the Zero.

One of these approaches, mostly a philosophical one, regards the Zero as an abstract concept and equates it to Nothing.

The concept of Nothing is then interpreted by some as absolute void and by others as "lack of something"

The mathematicians, after a long and tortuous process, came to view the Zero as a number in its own right, but in the end they offer the same interpretation of Trilussa: something that does not hold value in itself, but if it is associated with other numbers it drastically changes their meaning.

Therefore, an abstract tendency is to consider the Zero a concept, an idea. Another tendency is to express the Zero as a somehow concrete entity, while still equating it to "nothing".

Now, none of these approaches can satisfy the need to define the sense of wonder that the Zero is capable of creating; we therefore aim to introduce a third, maybe new way to conceive this fascinating, complex, mysterious and yet essential idea: the Zero as the universe. Full, amazing, immense, in a way inconceivable to our little minds.

But let's proceed with order: we should start with taking into consideration – only to quickly dismiss it – the idea of the Zero as Nothing.

3 ZERO AS NOTHING, NOTHING AS EMPTINESS

For now, let's not think of Zero as a number; let's think of it as a concept, without believing that the perception of Zero as a concept was born only after its definition as a number.

Some thinkers have always assimilated Zero to the idea of Nothing, a concept of Nothing which is structured in various expressions.

For example, mathematicians often use the term "Null" to identify an empty set.

Scientifically speaking, we can say that a set contains "null" if and only if it is an empty set.

In this sense, the cardinality of the empty set (meaning its dimension), is Zero.

The first modern set theory, developed by Cantor in the second half of the 19th century, was the focus of many debates between 1890 and 1930 and has been frequently "readjusted". In this time two main axiomatic systems were assessed: the *Axiomatic system of Zermelo-Fraenkel* and the *Axiomatic system of Von Neumann-Bernays-Gödel*. Each contains very specific axioms.

Anyway, as we have said, in set theory, the empty set is the one that does not contain any element. Another specific axiom was developed for this set, the axiom of empty set, which states:

There is a set X such that no set Y is a member of it

With yet another axiom, namely the axiom of extensionality, we can demonstrate that such set is unique.

The axiom of extensionality is among the most interesting axioms formulated in the 19th century by the German mathematicians Ernst Friedrich Ferdinand Zermelo and Adolf Abraham Halevi Fraenkel.

The latter in particular has made precious contributions to logic. He formulated two different hypotheses to anchor set theory to axiomatic foundations in order to avoid any sort of paradox, so as to improve the system of axioms developed by Zermelo.

The axiom of extensionality (or extension), also known as the axiom of Zermelo-Fraenkel, states that:

Given any set A and any set B, A is equal to B if and only if, given any other member c, c is a member of A if and only if C is a member of B. Essentially, this axiom states that two sets are equal if and only if their members are exactly the same. But in this case the given set A, which does not contain any member, because of the fact that it does not have any member C to compare to some other set, it cannot be equal to any other set, therefore it is unique.

Basically, if the empty set is unique, we might as well give it a specific name: we can call it "empty set". The finite sets – the ones that are not empty – are based on the empty set, but this is not the subject of our work.

We have considered the empty set. This set is often called "null" (meaning that it has no value), even if this might cause a misunderstanding in regard to the concept of a set that does not have any member.

The null set is identified mostly in the theory of measure, where it indicates a set which is negligible for measurement purposes.

We must then assess that the theory of measure is the branch of real and complex analysis which studies sigma-algebras (the families of subsets of a non-empty set Ω, omega, with closure propriety under some set operations).

Therefore, what the theory of measure studies are measurable spaces, measurable sets, measurable functions etc. The very notion of measure, and the concepts related to it, were developed between the 19th and 20th century, within the formalization of the theory of measure.

Back to our sets.

Here is how the empty set – the one that does not contain any member – is graphically represented. We can indicate the empty set with the symbols { } or Ø. The latter was used first by a group of mostly French mathematicians in the beginning of the 20th century.

These mathematicians employed the collective pseudonym "Nicolas Bourbaki".

Between 1935 and 1983, a series of books were published under this heteronym: the objective was the systematic exposition of notions related to modern advanced mathematics. With this scientific operation, the group aimed to base the entire mathematical science on the set theory.

Among these enlightened mathematicians there was André Weil, who was apparently responsible for the introduction of the symbol Ø to represent the empty set, in 1939.

Weil, does that ring a bell? Well, in fact, André's younger sister was Simone Weil, the extraordinary French philosopher and historian.

Simone, whose full name was Simone Adolphine, was born in Paris in 1909. She had quite an eclectic mindset and she studied mainly philosophy, literature and also mystique. She is well-known for her vast literary and scientific work, but also for the dramatic events in her life.

She chose to abandon her teaching career to experience the working class condition and, despite her persistent health problems, she was involved with social struggles, especially in regard to anarchy and Marxist heterodoxy.

So, these Weil siblings were extraordinarily gifted. André got into mathematics when he was only 10 and his personal contribution became fundamental to the progress of many branches of this discipline.

After having payed our respects to the Weils, we can get back to the empty set: we should try not to confuse the symbol Ø with the Greek letter Φ (*phi*) or the scandinavian vowel Ø, even if apparently André Weil chose the symbol because of the scandinavian letter.

The notation { } is perhaps a clearer and simpler way to identify the empty set.

We must also remember that the symbol {Ø} indicates the set that contains the empty set. This should obviously not be confused with the empty set Ø or with the digit Zero, which is often represented as Ø. More on Zero as a digit will follow.

There! We are now able to tell one symbol from another and if we happen to find all the notations Ø, Ø, Φ in the same text – admittedly, this is unlikely – we can easily tell the difference.

The symbol for the empty set, Ø, is based on a circle, like the letter O. The scandinavian letter is an oval with an oblique bar. The bar in the Greek letter Φ is straight, not oblique.

Alright, we have examined the Zero as Nothing and the Nothing as emptiness.

4 NOTHING AS NULL

The concept of *niente* (Italian word for "nothing") is often used as a synonym of *nulla*. In fact, the two terms are quite different. First, the etymology of the Italian word for Nothing, *niente*, is unclear and consequently it is basically impossible to be certain of its meaning.

The term may come from the Latin forms *nec entem* or *nec gentem*, which mean respectively "no entity" and "no people".

We often employ more complex locutions to avoid the uncertainty of the term. For example, instead of saying "there is nothing inside", we could legitimately say "there isn't anything inside", or we could definitely say "everything is present" instead of "nothing is lacking".

If we analyse the Italian term *nulla*, we notice that it is a noun, and nouns always represent a thing: the term *sostantivo* (Italian for "noun") comes from the

Latin word *substantia*, which means "substance", "reality". But can a *nulla*, which is a *niente*, have some substance?

Thus, Zero is a mystery, even in its grammatical definition.

Modern logic tries to assess the differences between these concepts and many philosophers argue that the term *nulla* cannot be considered a noun, because it represents nothing. But what is it then?

5 ZERO AS OPPOSED TO ONE

From time to time Zero has been equated to Nothing and as a non-entity, as opposed to One as an entity.

We are still discussing a concept, not a number.

For example, in computing, or better in machine programming, Zero is associated to a very specific state.

In order to transmit orders or instructions to a computer you need to write some programs. The language employed to do so, called machine code or machine language, is based on a binary alphabet. The processor or CPU is the hardware component of the PC that is able to run programs written expressly in machine language.

We should take into consideration that the tasks the computer is able to perform may be simple, but the computer is able to perform them in an infinite speed compared to us. Thanks to this speed, albeit just following instructions, the computer is able to do things which are unconceivable to us.

The easiest step, the initial brick, to perform every operation, even complex ones, is to detect the difference between two specific states, for example of a circuit: Zero = no current flows , 1 = current passes.

This is roughly what the machine language, based on binary code, expresses.

In computing, the definition binary code refers , in general , to notations that use binary symbols (0 and 1) or bit. The term is also used in a generic sense to mean a code (in the information theory, cryptography, or similar disciplines) that uses an alphabet composed of only two symbols: 0 and 1.

Starting from here, from the difference between 0 and 1, we were able to program the computer to send a man on the moon and more. Yes, in 1969, with the Apollo 11 mission, the Americans led the man to the moon, an impossible task without the help of computers, even if the computers of that time were less performing than a current digital wristwatch.

Let's review this quickly: in order to program machines a binary code (or machine language) is used, which is basically composed of only two symbols: 0 and 1. One indicates the flow of electric current (in a circuit), Zero indicates the lack of this impulse. The absence or the presence of something. Zero as opposed to 1.

Here's the news. This step brings us to the consideration of Zero as the lack of something. In this case the electric current, but the idea of Zero as "lack" is often interpreted as a shortage in an absolute sense.

A deeper research on the concept of Zero as a lack is then required; our analysis is shaping up to be something pretty interesting.

6 NOTHING AS LACK OF SOMETHING

We have surveyed Zero as Nothing and Nothing as Emptiness. We have then considered that the term Nothing may also signify the absence of something. And the absence of something reminds us of the thing we are missing.

We then need to try to better understand the essence of this nothingness.

The essence of the concept of Nothing has always been the subject of analysis and studies on behalf of philosophers and theologians. In order to define "Nothing", one follows an ontological approach, that is, one draws from the philosophy of being and not being.

In reality, with ontological acceptation, and in particular with philosopher Parmenides, Nothing becomes "not being".

Parmenides was born approximately in 515 B.C.E. (give or take a year, we do not have reliable sources) in Elea, in Magna Graecia, in particular in present-day Ascea, in Campania.

His *Poem on Nature* is considered the one and true beginning of the history of philosophy. This work is composed of a *Proem* and of a main body of two parts: *The way of Truth* and *The way of Opinion*. Following the *Proem*, the first part is mainly concerned with *Truth*, and then with *Being*.

Parmenides sustained that the events of the physical world are illusions and contrary to what is normally perceived, he affirmed the reality of Being. It is Being that is immutable, ever-present, finite, immortal, unique, homogeneous, unmovable, eternal.

In this poem, he presented the philosophical constructs through the narration of an imaginary journey towards a Goddess's dwelling.

She is Dike, the goddess of Justice, who with her sisters Eunomia, goddess of Legal Order, and Irene, goddess of Peace, has the duty of making mankind respect moral and juridical laws.

Let's just say her opinion is to be kept in mind. Now, we do not have evidence that Parmenides actually saw the goddess, but if the story is truthful, she surely gave him precious advice.

Dike showed Parmenides the way of opinion, which leads to the appearance and deceit, and then the way of truth, which leads to knowledge and to Being.

Parmenides claimed that *"being is, and cannot not be,*

and not-being is not, and cannot be".

More precisely, it goes like this:

... Well then I tell you - and listen carefully to this speech - which are the paths of research that are worth pondering upon: one is that 'it is' and that it is not possible for it not to be, and this is the path of Persuasion (indeed, Truth follows); the other is that 'it is not' and that it is necessary for it not to be, and I tell you that this is a completely inaccessible path: in fact you could not even be aware of what is not (as it is not possible), nor could you express it.

... In fact to think and to be are the same.

This claim – "to think and to be are the same" – was later appropriated by Descartes.

René Descartes, Latinised into Renatus Cartesuis, and Italianised into Renato Cartesio, was born in France, in La Haye en Touraine, today known as Descartes, on the 31st of March 1596. He is considered the founder of mathematics and modern philosophy. But to most people he is known for the phrase *cogito ergo sum*, which literally means *I think, therefore I am*.

There, Parmenides and Descartes expressed the unquestionable certainty that man has of himself, in quality of a thinking being.

The dialectic between being and not being appeals to Shakespeare, who expresses it through Hamlet. William Shakespeare is considered the most important author in the English language and his plays are translated into all major language in the world.

Written between 1600 and 1602, *The Tragedy of Hamlet, Prince of Denmark*, known simply as *Hamlet*, is among the most frequently performed plays. Hamlet's soliloquy *to be or not to be* (Act 3, scene 1) is the most famous passage and boasts innumerable re-interpretations on the stages of the world.

Back to Parmenides, for him, with or without Dike, Being is free from imperfections and identical in each part. Basically, it resembles a sphere, while nothingness is simply not being.

7 FROM PARMENIDES TO PLATO

We started with Parmenides, to then embark on a brief journey in the concept of nothingness in philosophy.

We highlight, however, that Parmenides' *not being* is not *not being there* just as opposed to a generic *being there*.

But given that *not being* is opposed to *being*, this idea of nothingness is already approaching the notion of *lacking*.

In fact, how can one say what *is not*?

One may only by being clear about *being*, the idea of something that *is*.

Plato took this one step further.

The philosopher Plato, born in Athens in 428 B.C.E., laid the foundations of western philosophy, together with his master Socrates and his student

Aristoteles.

For Plato *not being* was definitely the *not being* of something that *is*. Therefore, not only does it make sense, but it also has the same concreteness of something that is there and which one misses.

It has the same form, the same substance, of that something that exists and which *not being there* refers to.

He expressed this remarkably and unequivocally in his work *Sophist* where he presented nothingness as *the not being of something's being*, thus as the otherness of what is.

Sophist is a dialogue dedicated to ontological themes and is one of Plato's later works.

The philosopher tackled the theme of *not being* and committed a parricide against Parmenides. *Sophist*, in fact, with its deceitful dialogues, disguises *not being* as *being*. It commits an offence against Parmenides' warning: *That which is not you must not force into being.* Plato defined *not being* as a way of *being*, as otherness (being something other than something). All which is, and partakes in being, results in *not being* as well.

Even ideas are identical to themselves, but different from each other, since one is not the other.

Reality transcends, in that it manifests itself in a plurality of beings, of which one is not the other. *Being* is therefore a plurality, while *not being* is infinite.

So for Plato things *are* and at the same time they

are not, in their participation in being. Thus, with his formulation of Nothing he actually surpassed Parmenides.

His attitude towards Parmenides then becomes parricide.

In fact, the term parricide, in the cultural sense, is used to define the process of overcoming or corroborating a master by a pupil. Thus, Plato's parricide against Parmenides is commonly known as "Permenicide", or the ideal killing that Plato commits in his late years against Parmenides' thought.

It is also true that this path may lead to the mistake called reification. That is, one can mistake something that is not real for real.

Reification or not, postmodern existentialists, writers and philosophers embrace the notion of nothingness as absence of something, and not as lacking of everything, adhering with Plato's vision.

According to Plato, visible things around us are but imperfect copies of an ensemble of ideal forms. Models from which all objects derive their properties. These forms are eternal, immutable, and indestructible. Moreover, if all material being in the universe were to be eliminated, these forms would keep on being.

Then, if we know nothingness as one of these forms, it is impossible to conceive of an imperfect manifestation of this that does not deserve the name Nothing ("Nulla", as above).

But where are these perfect forms?

A void which contains even one thing is not a void. But the atomist model by Leucippus of Miletus and his beloved student Democritus is based on a void (if we may still call it so) that contains something. And here, if we talk about love, since we have mentioned the Athenian philosophers, we surely cause confusion.

In reality, we claim that Leucippus and Democritus' love is all but Platonic, and thus nobody shall feel their neurons are overcharged.

We do not have reliable information about Leucippus' life, who was born in Miletus, a Greek city in Asia Minor in the first half of the 5th century. The only reliable source is Aristoteles, who always mentions Leucippus together with his partner Democritus, with whom he funded atomism.

Democritus is Leucippus' student and partner, and it is practically impossible to distinguish his ideas from his master's.

Democritus was the most prolific writer among the pre-Socratic scholars, and is considered one of them, even though, in reality, he was born after Socrates, and died when Plato and Aristoteles were alive, possibly in his 100's.

Back to the atomists. In reality the philosophical model of atomism was already sketched in the end of the 7th century B.C.E, but it was affirmed with Leucippus and Democritus, so a few centuries later.

Possibly under the shadow of a Thracian walnut tree, Leucippus and Democritus spent their days pondering upon the essence of things. They might have been so caught up with comparing ideas between themselves and others, that they would even forget to feed themselves.

Anyway, they affirmed that both fullness and void are elements. Therefore, they considered one as *being* and the other as *not being*.

They identified fullness with being, void with not being. Being and not being have the same value. They mantained that *being* exists just as much as *not being*. Void is as real as fullness.

The atomists theorised that the natural world was made up of atoms, which is indivisible and can be joined, and of the void as a thing in itself and not a container of elements or things.

The legacy of the atomists is very important. It practically changed the world, the way we see it.

Now everything in in relation to other things.

Now we seem to perceive the indivisible of which we were unaware. A revolution, sure. But Aristoteles' vision is an equally precious legacy.

8 ARISTOTELES' VISION

Aristoteles was born in 384 B.C.E. in Stagira, modern day Olympia, which then was a Greek colony, in the north eastern part of the Chalkidiki peninsula in Thrace.

His visionary mind considered the natural phenomena, in particular seeking the finality in movement and change. Aristoteles' universe has a finite volume that encompasses every existing thing. It is full of matter and its space is defined by the bodies in it. On the other hand, nothingness has neither cause nor effect, neither reason nor purpose, and thus seems inconceivable in Aristotelian logic.

Centuries later, in Stuttgart, Germany, Georg Wilhelm Friedrich Hegel was born, who at the time was considered one of the greatest philosophers of contemporary history.

Hegel proposed a theory whereby the existing and the non-existing are equally undetermined as absolute concepts and are opposed to determined *being*. *Not being* influences the characterisation of *being*.

In Hegel's theory, becoming is very important, or the transition from *not being* to *being*. In this process, *not being* is transformed continuously into *being*. Therefore, in *becoming* we can say that *being* and *not being* are the same thing, given that the process is continuous.

If we imagine defining an object's life with a time line, then we define the change of the object from an instant to the next one as becoming.

In instant i the object is not yet what it will be in instant $i+1$, and so on. For this reason becoming is a constant passing from non-existent to existent, and then again non-existent in time.

Another German philosopher, namely Arthur Schopenhauer, who lived between the 700's and 800's, was inspired by elements of enlightenment, Platonic philosophy, romanticism, Kantism, and blended them with the oriental philosophy, in particular Buddhist and Hindu philosophy.

The result? Schopenhauer created his very own conception of philosophy, characterised mainly by a strong pessimism. This conception has had an extraordinary influence – although sometimes re-evaluated – on following philosophers, such as Nietzsche, and in general on European culture of his time and following ones, and his is part of what we call the philosophy of life.

This new philosophical field was developed around the end of the 19th century in opposition to enlightenment, positivism, and intellectualism, and is often defined with the term irrationalism. The first elements are found in German romanticism, which conceived the existence as a continuous tension of finiteness towards infiniteness.

Back to Schopenhauer, we saw that for him Being is given by the objectivation of Will, while Nothing is the result of the negation of Will itself.

We cannot say anything about Nothing, since our perceptive and conceptual systems are born by Will, and serve its purposes. Nonetheless, Nothing is humanly reachable and attainable, as it happens in the ecstatic experiences of mystics.

9 NOTHING AND NIHILISM

Nihilism, from Latin *nihil* and medieval *nichil*, that is nothing, is the doctrine that purports the negation of one or more aspects apparently significant to life.

From here the world (and human existence in particular) is empty of meaning, purpose, ethical value, and truth is incomprehensible. In the interpretation of existential nihilism, life is without meaning, without objective and without intrinsic value.

One thinks of Nietzsche, even though Socrates may be thought of as a father of nihilism.

Friedrich Wilhem Nietzsche was born in Saxony-Anhalt, in Germany, in 1844.

He is considered to be among the major philosophers and prose writers of all time. He has had a controversial, yet unquestionable, influence on philosophical, literary, political and scientific thought in the 900s. His philosophy is part of the philosophy of life and is considered by some as the boundary between traditional philosophy and a new mode of reflection, which is informal and provocative. Nihilism states that morality does not exist for itself, and that all values are established abstractly and artificially.

The term nihilism is sometimes used in association with anomia, a mental state of despair and of perceiving existence as pointless.

Schopenhauer, Nietzsche and many other philosopher embrace negation to dive into Nothing. Even Leopardi does so, or better count Giacomo Taldegardo Francesco di Sales Saverio Pietro Leopardi.

Leopardi was born in Recanati, in the Marche region, in 1798. He is considered the most prominent poet of the 800s in Italy, and one of the most important figures of literature in the world, as well as one of the main protagonists of literary romanticism. Leopardi, much like Schopenhauer, Kierkegaard, Nietzsche, and later Kant, is seen as an existentialist or at least a precursor of Existentialism.

In the poem *To self* he wrote:

Bitterness and boredom is life, nothing ever else; and mud is this world.

We have begun to see that the perception of Nothing has been articulate in time and in several cultures. Let's keep on doing so.

In Sanskrit, the Indian language in the Indo-European family, the term *"shunyata"* means Nothing, absence, and this word is often mentioned in Buddhism.

Let's remember that Buddhism is one of the oldest religions, or rather philosophies of life. It was founded by the wandering ascetic Siddharta Gautama, who lived between the 6th and 5th century B.C.E.

Buddhism represents a mix of traditions, thought systems, practices and spiritual techniques that come out of the different interpretations of life practices. Starting from India, Buddhism expanded in the following centuries mostly in south east Asia and the Far East, arriving in the West in the 19th century.

Back to the term, *shunyata*, it is from this term that some scholars establish a link between Buddhism and existentialism.

Nothingness in Buddhism established what is and is not.

Then there is *dukkha*, another Sanskrit word, which expresses sufferance, which comes about from the awareness of that nothingness.

There is the unborn, the unoriginated, the uncreated, and the unconditional.

If it wasn't for these, there wouldn't be a way out of the born, the originated, the created, and the conditional. But since the unborn, the unoriginated, the uncreated, and the unconditional exist, there is a way out.

This is what Buddha said. This is what makes us

seek something that provides relief, that frees mankind from perennial *dukkha.*

We must say that in some forms of Buddhism Nothing is not considered as not being, but rather as a mental state (nirvana). In this sense who reaches Nothingness is capable of being completely concentrated on a thought or activity. To such a degree that would not have been possible if one were consciously active in thought. So to be active, paradoxically, one must reach Nothingness.

But then, is this not sufficient to show that nothingness is indeed everything? We will explore this later.

10 THE TRAGIC NOTHINGNESS OF EXISTENTIALISTS

In the West, existentialism is founded on the specific value of the individual. On their precarious and finite character, and on the loneliness in front of death.

Existentialism as school of thought is expressed in philosophy, literature, the arts, and costumes.

It originated between the 18th and 19th century, even though it was established in the 20th century, spreading more decisively between 1920 and 1950.

The founders are Germans Karl Theodor Jaspers, born in 1883, who was born in the small town of Oldenburg, in lower Saxony, and Martin Heidegger, who was born in Meßkirch, in Baden-Wuettermberg, a few years later.

Jasper was a philosopher and psychiatrist. He decisively contributed to philosophy, psychiatry, but also to theology and politics.

Heidegger is held as the main representative of ontological and phenomenological existentialism.

The philosophers of existence, starting from the two founders, showed a complete awareness of the drama of the world's being.

For them, to be is tragic, without a chance of liberation.

Here is then the nothingness where mankind is devoured. But while the Buddhists call this suffering *dukkha*, envisioning a way to overcome or at least put up with it, the West coins the term anxiety. So if dukkha meant hard to bear, for the existentialists this weight is practically impossible to bear and degenerated into anxiety.

How would we call this misery nowadays? Maybe just depression. That state of (wrong) awareness of being alone in the world, of having nothing but void around, a void that is paralysing, and thwarts all attempts at entertainment, or any action.

Quasimodo wrote in the poem *Ed è subito sera* (*And suddenly it is night*):

Ognuno sta solo sul cuor della terra,
trafitto da un raggio di sole:
ed è subito sera.

English version

Everyone is alone on the earth's heart,
Pierced by a ray of sunshine:
And suddenly it is night.

But if we live wrapped in void, that is if void engulfs us, can we call it void?

And we do not wish to bring up religion or faith, two concepts that are always to be considered separately, otherwise it is easy – too easy – to demonstrate how the feeling of being alone in the world is unfounded if one is aware of God's existence. God, not religion. God, whichever name one gives to it.

11 FROM PHYSICS TO METAPHYSICS

Well then, to conclude the exposition of Zero as a stat and not a number, the concept of nothingness is articulated from physics to metaphysics, to arrive at religion.

In physics, the word Nothing is not employed in any technical way, although we consider Nothing as empty.

In physics, we define emptiness as a subset of space that does not contain any matter, but that may contain physical fields (electric fields, gravitational fields, etc.). It is a subset undergoing some kind of force.

It is impossible to have a subset of space without matter nor fields, since gravity cannot be blocked and all objects are subject to electromagnetic forces.

But here we enter the field of quantum mechanics, which covers the world and the universe starting from quantums, which are the microscopic concentrations of energy composing matter.

In the blink of an eye, quantum mechanics supersedes classical mechanics.

Yes, because in the 800s it was thought that the principles of nature were solved. Atoms are the bricks that make up the world. Newton and Galileo's works seem sufficient to explain the movement of planets and all other bodies, and no other scientist seems to have such visionary epiphanies.

But in the second half of the century, some incongruences between theory and experimental data arise. It is the crisis of classical physics, since new observations call into question classical assumptions.

So, in the 20th century, several scientists again investigate the nature of things. The doubts from previous generations find answers in the theory of relativity, and quantum theory. In fact, a revolution occurs in scientific theory, marking the shift from classical physics to modern physics.

In particular, the German physicist Max Planck introduced the concept of discrete energy, and called quantum the value of energy allowed.

Planck asserted that some measure of certain physical systems only vary by discrete – not continuous – amounts, i.e. by a quantum.

Max Planck was born in Kiel, Germany, in 1858.

In 1900 he presented the hypothesis that energy shifts in emission and absorption in electromagnetic radiations occur in discrete steps, not continuously, as sustained by classical electromagnetic theory.

In 1901 he went from hypothesising to theorising this notion of discrete quantum energy increases. In this way, energy can be represented granularly, like matter. Quantums are indeed granules of energy.

This theory got him the Nobel prize in 1918. Plank was very religious. On religion and science, he wrote:

Science and religion are not in opposition, but do need each other to complement one another in the mind of a man who thinks seriously.

Another merit of quantum physics is the gradual understanding that our knowledge of reality is far from complete.

Quantum physics revolutionised the concept of measure.
Measuring a size is actually finding a relation between it and another one that is homogenous to it. It mainly requires two fundamental elements: a measuring system, consisting of tools, and a methodology that allows to utilise such tools.

Quantum physics, in particular the principle of

non-determination, establishes the impossibility of knowing the state of a particle without irreparably perturbing it.

Even if we can imagine it possible to measure a quantity of a system, one cannot determine what its value was before the measurement took place.

Absolute zero, -273.15C degrees, is officially the lowest reachable temperature, in which particles stop moving.

Again, practice confutes theory. Recently some scientists managed to go below absolute zero.

Do we need to re-evaluate the laws of physics? Maybe.

Back to our topic, the notion of nothing as emptiness, even if we assume it possible to have a space with no matter nor fields, this would be circumscribed in an unmeasurable way.

Here we touch upon the quantum mechanic emptiness theorised in 1976 by Italian physicist Massimo Corbucci, who proposed a new system of atomic levels.

Without diving into the details of quantum physics, we must ask: can something measurable be called nothing?

In going from physics to metaphysics (i.e. beyond physics), beyond the mutable, unstable, accidental phenomena of physics, one focuses on the eternal, which is stable, necessary, absolute. One seeks this to find the essence of being.

The relation between ontology and metaphysics, as we have shown, is the study of being in itself, and becomes quite tight.

As we saw, Parmenides was the first to ask the fundamental question: why being and not nothing?

But notions about God, the soul's immortality, being in itself, origin, and the sense of universe belong to this metaphysical research, as do the relation between the transcendence of being and the immanence of material things.

12 NOTHINGNESS IN RELIGION

In Western philosophy, the notion of nothing is not only used in opposition to being, but also in a theological sense, as the non-divine (as intended by Kant and Descartes), and as profound essence of the divine, as by the philosopher Eriugena, the theologian Meister Eckhart, and Jacob Böhme.

The last of these is one of the most prominent representatives of modern Christian mysticism.

As non-divine and non-spiritual, nothingness is often identified with matter. This is the viewpoint of Greek philosopher Plotinus and Augustine of Hippo.

So here we get to religion.

Many scientists look for evidence of the absence of God, and some arrogantly claim to have found some, or at least they seek evidence that nothing justifies his existence.

Stephen Hawking, the brilliant physicist, mathematician, cosmologist, and astrophysicist, is known for his studies on black holes.

Hawking, among others, partook in the elaboration of many theories: the multiverse, the formation and evolution of the galaxy, and cosmic inflation.

Like Newton, the scientist is a fellow at the University of Cambridge teaching mathematics for about 30 years, from 1979 to the 30th of September 2009. Hawking is a member of the Royal Society, of the Royal Society of Arts, of the Pontifical Academy of Science.

In 2009 he has received the presidential Medal of liberty, one of the highest badges of honour in the US, conferred to him directly by president Obama.

So, he is a man and scientist of big calibre. He proposes that there may be a unifying theory of the universe, and in substance brings back Einstein's ambition explaining physics with one formula, finding the mathematical secret at the heart of the universe, and reunifying the forces of the infinitely small and the infinitely large.

Stephen Hawking has created the theory of

Everything (TOE), which explains how to link together all physical phenomena.

He is an atheist and his research is thus oriented, as we'll see. He writes:

The universe may create itself and it creates itself from nothing. Spontaneous creation is the reason why there is something instead of nothing, why we exist. There is no reason to appeal to God.

Hawking is basically claiming that the universe comes from nothing, but he does not explain exactly what this nothingness is.

Is that a simple not being or rather the not being of something that exists?

The difference, as we have explained before, is remarkable.

Another famous mathematician who tackled nothingness and ended up in theology is the Italian Piergiorgio Odifreddi.

Odifreddi, born in 1950, is a mathematician, logician and writer and is met with perplexity when he moves from philosophy to theology to relate Zero (which he interprets as nothing) with God.

Sure, all opinions may be listened to, shared, and maybe criticised, what shocks is that people who claim to be atheists want to dabble in intricate formulae, theories and hypotheses and other scientific tools to prove or disprove the existence of God.

With his claims in the text titled *Dear Pope I'm writing to you*, he exposes the contrariety of Pope Benedict XVI, who even replies to him. A very unusual reaction. Pope Ratzinger (Benedict), is a man of culture, with a open mind, and enlightened. He accepts any expression as long as it is critical and founded.

He is annoyed by claims that are motivated not by critical reasoning, but rather by... nothing, indeed.

The pope has expressed his disapproval of Odifreddi's claims regarding Jesus and theology in general.

We are now aiming to highlight the perplexity over Odifreddi's perception of zero.
He uses terms like nothing, empty, zero, as if they were synonyms.
First of all, he speaks of zero while actually speaking of nothingness, but then, why not use a more apt term? Nothing, in fact. Without picking on zero, which is so dear to us.

Odifreddi claims to be an atheist with scientific complacency, but also feels entitled to express himself on the concept of God. Shall we think of him as someone who, like all human beings, sometimes makes a mistake (a cultural-social one, in this case)?

From his texts it seems he is a little confused.

With the benefit of the doubt. But since he knows the meaning of words, of all words, why not choose a more appropriate one? Thus the doubt over his confusion is legitimate.

Odifreddi claims there are many ways to approach the notion of zero-nothing-empty. But seems to consider the terms in the same way, and shifts from religious silence to universal hole to existentialist emptiness.

He says for example that at times there is nothing else, as in the musical piece titled *4 minutes 33 seconds* by American composer John Cage, whose work in central to the evolution of modern music.

In *4 minutes 33 seconds*, he expresses 273 seconds of silence, which represent the absolute zero temperature. Silence without a note. So here zero is intended as nothing.
Odifreddi is not the only one conflating these notions.

Even Simone Weil, who we mentioned earlier in

relation to her brother Andre, who introduced the symbol for empty sets, even she addressed the issue of nothing.

She wrote:

There is nothing from which we run, and there is nothing towards which we move.

She conceived nothing as emptiness, as not being, and thus claimed we run from nothing at birth and come back to it in death.

For a woman who left teaching to experience the working class condition, and then to become an active dissident, inertia, giving up, are already part of nothingness, of not being.

Many other have addressed this topic, but almost all express nothingness not as emptiness, but rather as everything, as fullness.

We approach the idea that nothingness is not nothing, but rather a sort of container.

Basilides, a Greek religious master who lived in the second century AD, laid the foundations for Christian Gnosticism.

He was born in the great Alexandria of Egypt, on an unknown day. He was a Roman army man and is venerated as a saint by the Catholic church. It is thought that he wrote several commentaries on the gospels, which are now lost.

His supporters formed a movement that lasted until two centuries later. His school then fused with mainstream Gnosticism.

Interestingly, he wrote:

The nothing-God created from nothing the nothing-world

Why interesting? Regardless of one's faith or lack thereof, can one express God as equivalent to nothing?

If he was nothing, what need to give nothing a name?

John Scotus Eriugena was a monk, a philosopher and theologian, who was born in Ireland in 810. He was also involved in the path to the unification of zero, nothing and emptiness.

He claims:

Nothing from which God creates all things is God itself.

Even in this case, it is clear that this nothingness is fullness, the everything from which all is derived. According to Christian religion, God has created all things from nothing.

Augustine of Hippo (354-430 A.C.), one of the

main thinkers of the Catholic Church, claimed finiteness is a symptom of being derived from nothing.

Somehow in creation creatures are found between Nothing and God. Can we therefore hypothesize the equation Nothing = God?

13 THE WORLD FROM NOTHING

The Lutheran philosopher Jacob Boehme was born in Alt Seidenberg, in Saxony, in 1575. He is considered one of the main representatives of modern Christian mysticism, and is named Philosophus Teutonicus by his coevals.

Boehme reflected on the relation between God and Nothing. In particular he pondered creation.

As Christian religion does, he also highlighted that God created the universe from Nothing, and that before creation nothing existed, not even God. But then, how was God created?

In particular Boehme claims:

Nothing is God made all things from Nothing, and he himself is Nothing.

But this nothing is a weird nothing. It is not at all nothing. Then what is it? God itself is seeing and feeling nothing. And he is called "Nothing" (despite being God) because at the end of the day he is utterly incomprehensible.

So this nothing is in fact fullness. It reminds us of Michelangelo's idea that his sculpture is already in the marble, one simply needs to set it free.

Back to the mathematician Odifreddi, in his mix-up between emptiness, nothing and zero, he also brings in quantum mechanics, but again what he calls zero is nothingness and what he calls nothing is everything.

We have in fact seen that in quantum mechanics emptiness is a region in space where in reality many things happen.

Matter and antimatter is produced, of duration inversely proportional to their mass. It was the German scientist Werner Heisenberg who in 1927 posited that matter is created from the eternal nothing, by the famous principle of non-determination.

He said that nature borrows energy, for a period as short as the amount lent is large.

He established the limits of knowledge, or determination, of the values of physical measures associated with operators that do not commute among each other, and which take on a physical system at the same time.

It seems then that for modern physics emptiness is the natural cradle of existence.

We thank Odifreddi for his diligent work, but we

may say again that it is relative to nothingness, not to zero. We may grant the acceptation of nothing as emptiness, but it ends there.

Zero is a whole different thing!

14 A MORE CONCRETE NOTHINGNESS

We saw zero conceived as nothing, and travelled rapidly through the various ways of perceiving nothingness.

Some called it a lack of something that exists, while others called it something that does not exists at all, nor can be defined.

Nothing as something that is not, and is boundless, is an undefinable and abstract concept.

Nothing as lack of something we can imagine, even abstractly, has a precise boundary, the same as that of the things we know and whose absence we perceive.

It's a bit like pushing an object against the snow or the sand. The shape remains (at least in part), but the object is gone. We perceive its presence. So nothing as absence is more concrete than nothing as nothingness.

This notion will come in handy when talking about numbers.

Let's take a step back.

We saw that in set theory nothing is intended as absence: the empty set contains no element.

We also saw that the axiom of empty sets postulates its existence. From this all finite sets are derived. The empty set is also called nought, but this may cause confusion with nought set.

In measuring theory, an empty set is one that is negligible for the measurement.

So in set theory, nothing does not have a technical meaning. It may also be said that a set containing nothing if and only if it is truly empty, then has cardinality (or dimensionality) zero. In other words, nothing is another way to refer to an empty set.

We are here using zero as absence. The empty set is for our neurons a bit like the shape in the snow. How could we otherwise conceive, without going crazy, of a set (which intuitive is bounded) that contains nothing, that is empty, and has no limits?

To summarise, we saw zero as a concept. Often we say zero to mean nothing. We then saw how nothing is intended as empty, not absolutely, and then as absence of something we conceive. That is, nothing as absence of something relevant (otherwise we wouldn't miss it).

Therefore, nothing actually indicates something, an event or object of significance.

15 THE MATHEMATICAL STANDPOINT

We finally get to the concept of zero in mathematics. But it is necessary to face the planet of numbers in general before getting to the specific definition of that special, mysterious number that is zero.

That's right, because after such a long way we finally got to a very interesting field. The one in which zero is considered as a number. And a particular one indeed. It symbolises what is before one, this is clear.

But it is harder to conceive that zero is not only before one, but somehow it also contains it, indeed one to the power of zero is equal to one.

We'll expand shortly.

For now let's just take a small step. Let's associate

zero to the perception of something. Zero is something that is more than nothing.

It is the inception. The inception of everything.

It is what is before it is. But do we interpret it as the not being before something is, or rather as something that allows for what will be?

It indicates the beginning of numbers, the beginning of time, and the inertial state of the beginning that has no time nor space, no matter nor dimensions, no consistence, and yet represents possibility, and indeed necessity, that everything may be.

We saw then a concrete zero, an initial position which is tangible and measurable.

On the other hand, nothingness identifies…nothing, and more importantly it has nothing to do with facts, but only the sense or nonsense of things, their being and not being.

OK, so we are handling something more interesting. Let's start to understand zero no longer as nothingness or lack of something, but as a number.

There, in moving onto a more concrete perception of zero, in attributing it the weight of a number, the analysis seems more interesting, although, as we'll see, it is still far from sufficient to understand, encompass, and engulf the true essence of zero.

Zero is called صفر (sifr) in Arabic, אפס (éfes) in Hebrew, शून्य (śūnya) in Sanskrit. The Greeks called it μηδέν, but unfortunately this is still intended as nothing.

Here and now, it is important to distinguish the concept of zero as absence of value from that of zero as a number.

Let's give a practical example with water and temperature.

Before this, let's remember that there are two scales for temperatures. The one most used in the US is Fahrenheit, while in Europe we use Celsius. The latter is owed to Swedish physicist Anders Celsius, who in 1742 presented a memoir to the Swedish Royal Academy of Sciences in which he proposed a temperature scale where the unit of measure is named degree Celsius in his honour.
Then, we get back to water and decide to utilise the Celsius scale.

Let's say that if the temperature is zero, water freezes (according to the Celsius scale), and if temperature as data is missing, that is if we have absence of a value, we may not say anything.

There, this small example tells us the difference between thinking of zero as a number and picturing it as absence.

So to think of zero as a number leads to some meaningful considerations. To understand, we must start from the history of numbers.

First, we ask when the notion of numbers arose.

We can think of this history to be as old as man's history itself.

All civilisations, even the most primitive, have some notion of numbers. Many archaeological findings have confirmed this observation, such as bones of animals with different marks. This tells us primitive men used to mark objects to count. They counted, for example, animals, spears, or maybe women, since polygamy was adopted and accepted to preserve the species.

So since the beginning of time mankind felt the need to count and, unconsciously, to be "one that counts". Look at that!

With the spread of the need to count, several counting systems also rose.

The oldest? It is told, perhaps, by a wolf's bone dating back to 30,000 years ago.

The bone was found in Dolni Vestonice, a borough in the Czech Republic, which is part of Breclav, in the region of southern Moravia. Today it is known for this archaeological site where pre-Neolithic remains are found.

The bone in question was found by chance in

1937 and it displays 55 marks distributed in groups. We can say it is a witness to the most ancient computation which made it to our day. It tells us that the calculation was performed by grouping numbers by five. Favouring five (possibly because it matches the number of fingers) results in a system in base five.

But here, as well as a counting system, we are seeing numbers too, which we almost take for granted.

But who invented numbers, those little symbols that replaced the engraved marks (on a bone or elsewhere), thus facilitating far more important computations?

Let's take a step back.
All people, in all parts of the world have developed the notion of numbers.
Analysing the various counting systems, it appears clear that the idea of such a system occurs to many peoples at the same time.

The first were the Indians (as in the Indians of India, not America).
Arabs had then the merit of spreading numbers to the West thanks to the work of mathematician Muhammad Ibn Musa al-Khwarizmi.

Al-Khwarizmi was born in Baghdad, then Persia,

in 780, he was a mathematician, astronomer, and astrologist. In Baghdad he lived under caliph al-Ma'mun, who nominated him responsible for his library, the famous Bayt al-Hikma, or *Home of Knowledge*. Under his direction, several of the main mathematical works were translated into Arabic from Greek, coming from ancient Persia, Babylonia, and India.

So we saw that the notion of number was born and spread among continents without problems. But what about zero? Zero was late in its descent onto mathematicians' minds.

Egyptians were definitely the true masters of geometry.

Greek philosopher Plutarch was born in Cheronea in 46 A.C., and lived under the Roman empire for which he carried out administrative roles. He studied in Athens and was strongly influenced by Plato's philosophy. He wrote the biographies of the most famous people of his age. He told us of Thales, born in Miletus in 640 B.C.E., and is considered the first philosopher of history in Western thought since Aristoteles.

He told us of Pythagoras, philosopher and mathematician, who was born in Samos in 470 B.C.E. and is remembered as the founder of a school where he developed his mathematical theories and applications, among which Pythagoras' theorem.

Thales and Pythagoras, according to Plutarch, learned geometry thanks to the Egyptians.

Several Egyptian papyri have been found, and are

testimonies to the knowledge of the time. Then, Egyptians knew how to measure land and calculate the borders of fields after the floods from the river Nile. They knew the formulae to calculate the area of plane figure and the volume of solids such as the pyramid. Yet in the papyri there is no trace of zero. Greek mathematics contributes considerably to this with the creation of logic, i.e. the branch of mathematics that studies formal systems from the point of view of the codification of intuitive concepts of demonstration and computation, as part of mathematics' foundations.

This innovation paves the way to what later became modern mathematics, yet even here, in the beginning, there was no trace of our wanted zero.

16 COUNTING SYSTEMS

So, starting from the beginning of the history of numbers, zero was not really missed (as a number, clearly).

After all, our journey started from additive systems. Numbers are summed, maybe by grouping them by 5 or even 10, and there is no need for this mysterious element.

In mathematics, an additive counting system is one based on the additive law applied to specific fundamental numeric symbols. Each number is represented by a sequence of such symbols and its value is given by the sum of the values attributed to each symbol. Thus, in practice, zero is unnecessary.

For example, Egyptian counting uses repetition of a sequence of symbols representing one, ten, a hundred, one thousand, ten thousand, a hundred thousand, a million.

The signs appear in increasing order of size, even though this was simply a stylistic choice.

In numerals, the positions of the symbols relative to each other is not relevant to convey information, since each number has its very own symbol, or is the pairing of several symbols.

In practice, if numbers can be in any position without modifying the real amount they represent, there is no possibility for an empty slot or a sign that may represent this, since this would not make any sense.

Sumerians were the oldest known stationary people. They lived in Mesopotamia, modern day Iraq, from 4500 B.C.E., although this is argued.

Well, it was them who attempted to solve the problem introducing for the first time a new element in the counting system.

In all of Babylonian civilization (that is all the people succeeding each other in Mesopotamia, such as the Babylonians, Assyrians, etc.), the same not purely decimal system was used, which is base ten, but also which introduced 60 as a base. The symbols represent one, two, ten, sixty, six hundred, 3,600, and 36,600.

Then, according to a pre-established scheme, a multiplicative notation was developed. There were fewer symbols and the big numbers had an internal logic allowing to generate larger numbers with smaller ones, without having to create new symbols.

The numbers were pressed into clay boards. A significant change occurred in 2600 B.C.E., thanks to the Babylonians, with the invention of a stylus capable of producing thinner lines of different dimensions.

Then, a positional system appeared in Babylonia in 2000 B.C.E. It was limited to extending the cuneiform notation and the old additive system in base sixty so as to include positional information.

At first, this last method was used in particular by mathematicians and astronomers.

It soon spread to accounting in the royal courts.

In the sexagesimal system, the presence of an empty space makes a difference.

Using only two symbols, one for unity and one for ten, these people, in particular the Sumerians, were capable of writing any whole or fractional number.

Seems complicated?

Let's take an example to show how simple it is to count with this system.

Let's take the number 424,000.

The Babylonians wrote this conceptually as

1,57,46,40. The comma was used to separate sexagesimal numbers.

This notation means:
$1(60^3) + 57(60^2) + 46(60^1) + 40 = 424,000$

The Babylonians managed this with only two symbols -- for one and ten -- other than empty space (zero).

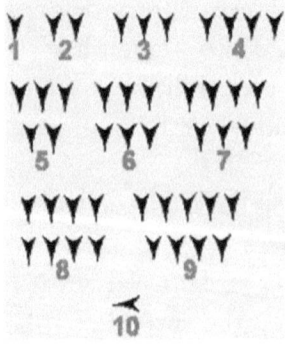

Source: Wikipedia

The number is:

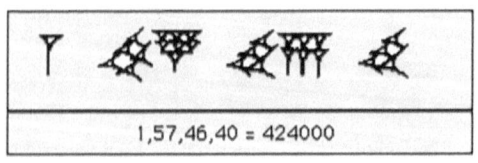

Source: Wikipedia

The representation of the number multiplied by a power of 60 is expressed additively, so each number constituting a multiplier for the power of 60 is represented as a sum of its units.

In the example, 57, which multiplies 60^2 is formed by five times the symbol for 10 and 7 times the symbol for 1. It is therefore a mix of addition and multiplication.

17 A WORLD WITHOUT THE ZERO?

In the Babylonian counting system, one must be mindful of spaces to avoid confusion. When there is more than one space, it is hard to evaluate the number.

This is the reason why, 1500 later the symbol for zero came about, introduced by the Babylonians.

This sign consists of two overlapping wedges, a double cusp that allows to indicate an empty spot in the number.

This system was immediately adopted in astronomy, a science which was greatly important to the Babylonians.

This lays the foundations for the first symbolic representation of zero in human history.

Nonetheless, the Babylonia zero is simply an empty space. It is never used in operations, let alone associated with a metaphysical notion of nothing. Basically, it is a mere place holder. It meets computational needs, but also practical ones.

We still have the legacy of the sexagesimal system.

It is used in time measurements. If we want to represent 3 hours, 40 minutes and 20 seconds, we can write $3(60^2) + 40(60) + 20 = 13.220$ seconds.

We do not wish to explain why this system was adopted, as we would have to reiterate the history of numbers from its creation, which would be too complicated.

Instead let's get back to other systems.

The main one used nowadays is base ten and this is definitely less cumbersome than 60.

The oldest decimal system appeared in the 3rd century B.C.E., when the Chinese introduced positional value in their system. But they did not use zero.

The position requiring zero is simply left empty. In fact, calculations are carried out with sticks representing numbers and placed on a board.

The Arabic system (originating in India) is the most commonly used in the world. The first digit from right represents units, the second teens, the third hundreds, and so on. For example, 554 represents 4 units, 5 teens, and 5 hundreds. In a non-positional system, on the other hand, these three notions would be expressed through different symbols; such as in Roman numbers, where 554 is written DLIV.

The Mayans were found in Central America from the 1500th century B.C.E. This people also developed a sophisticated computation system.

In their classical ages (from 250 A.C), the Mayans developed a system in base 20, positional and including zero.

The numbers are represented with three symbols: an empty shell, a dot, and a line. The symbols represent zero, one and five respectively.

They are ordered vertically.

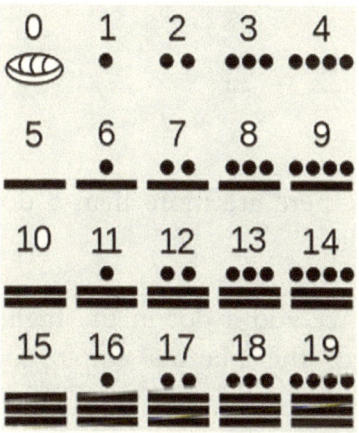

The first 19 numbers are made up of dots and lines according to an additive method.

When numbers equal or higher than 20 are written, a tower of symbols is constructed, the lower level representing multiples of 1, the second level multiples of 20 and so on. A bit hard to explain, admittedly. It is simpler to demonstrate it.

For example, 69 is:

●●● $3 \times 20 = 60$

●●●● $9 \times 1 = 9$

While a sum is defined by combining the symbols:

$$\underline{5} + \underline{\overset{\bullet\bullet\bullet}{8}} = \underline{\underline{\overset{\bullet\bullet\bullet}{13}}} \quad \underline{\underline{\overset{\bullet\bullet\bullet}{13}}} - \underline{5} = \underline{\overset{\bullet\bullet\bullet}{8}}$$

If after and addition there are more than 5 dots, we replace these with a line.

If we reach 4 lines, we add a dot in the highest row. As we mentioned, the Mayans employed a specific symbol for zero.

The function of zero is thus to indicate the absence of a multiplier in one of a tower's levels.

Zero is used in middle of final position.
Here is an example. In the table below two numbers are shown (read from top to bottom).
For 560 we use zero in the lowest level to indicate the absence of units.

Posizionamento delle cifre nel sistema maya		Esempi di numeri		
—	Cinque	8000 u		—
...	Tre	400 u	•	•••
..	Due	20 u	•••	☰
⊙	zero	1 u	⬭	••
			560	41.527

Source: Wikipedia

18 THE DECIMAL POSITIONAL SYSTEM

In our system, zero has a specific function, given that adding to the right side of a number means we multiply by the base value.

Then, to summarise, with the introduction of numbers, different people also thought of how to utilise them to make computations in a simple and fast manner. Hence several systems developed. The most common one in base ten.

Probably because we have ten fingers and thus it results simpler to count by looking at one's hands, even though some people use base 5 (especially earlier ones, as the one from the wolf bone).

Oddly, early settlers in France used a system in base 20 (maybe they used both fingers and toes to count?), which is proven today by the French for 80, which is quatre-vingts, literally four twenty.

The Babylonians, as we mentioned, used a system in base 60.

Today, in informatics we use a binary system (base 2), but also an octave (base 8), and hexadecimal (base 16).

Each people also has their own way of writing numbers.

The Romans used alphabet letters as numbers. So I meant 1, X meant 10, L meant 50, C meant 100, D meant 500, and M meant 1000.

The brilliant Romans are quite disappointing in this regard. It is pretty cumbersome to carry out operations using their system. It is not possible to arrange the numbers in columns to do addition or subtraction.

Indeed, both they and other peoples who used such systems had to use abacuses.

Anyway, the Roman system, like the Greek one, and the Egyptian one, is additive.

This means that each symbol that is added to the number represents a sum.

For example, if X represents 10, double X means

the number 20.

The system of the ancient Romans is yet not extinct, since it is still used today to represent ordinal numbers (I first, II second, and so on).

19 ZERO BECOMES A NUMBER

We have struggled a bit in reiterating the steps of the history of numbers and counting systems, but the positional system brought us to the notion of zero, with actual specificity.

Starting from the 7th century A.C., a new method for writing numbers with only 10 symbols was developed among Indians and then Arabs, such that each symbol has a different meaning according to its position.

In this system, Zero mathematically allows to introduce negative numbers and gives solid foundations to arithmetic.

The first study of zero is owed to mathematician Brahmagupta in 628 A.C.

Brahmagupts was in charge of the astronomic

observatory in Ujjain, India, and during his stay he wrote two works of mathematics and astronomy: the *Brahmasphuta Siddhānta*, in 628, and *Khandakhadyaka*, in 665.

Brahmasphuta Siddhānta is the most ancient known source other than the Mayan system, which handles zero as a number. It goes beyond. It sets the rules of arithmetic regarding negative numbers and zero, which are fairly close to modern day principles.

In *Brahmasphuta Siddhānta,* an arithmetic that included zero and negative numbers was proposed.

Brahmagupts delineated a numeric system made up of 9 digits (1 to 9) and a symbol, zero. With this system it is possible to write any number simply and precisely.

Only later Indian mathematics made zero equivalent to the other 9 symbols and came to see it a number as such.

Several debates are held around this concept, that zero is a number like others.

Let's consider for example Mahavira, the Indian mathematician. He was born in Vaishali in 599 A.C. and was one of the greatest Indian spiritual masters, and a contemporary of Buddha. He was receptive to the sense of precariousness that the doctrine of samsara implies and pursues ascetic life. After long meditations and corporal flagellations he reached a doctrine that allows to overcome the cycle of existence. To become a Jina (winner) over passions.

But in this context he is very important to us, since he defined the rules of the four operations involving zero.

20 OPERATIONS WITH ZERO

Mathematician Mahavira stated for example that:

A number multiplied by zero is zero, and such a number is unchanged when divided by, added to or subtracted to zero.

Here we enter the glorious history of the number zero.

But according to Mahavira a number divided by zero is unchanged. Clearly he thought, for example, that if we divide something among zero people, that thing does not diminish. But then, again, Mahavira had an idea of zero that is closer to nothing than to a number as such.

Dividing 0 by a number, other than 0, is simple: the result is 0.

But what about the opposite? What if we divide a number by zero?

It almost seems meaningless.

How can we divide a number by zero? Then again, what about 0 divided by 0?

Now, there is no shortage of answers, in fact any number can be the result. For example, 0/0=7 or 9 or 20. The proof is that 7x0=0, but also that 9x0=0 and 20x0=0 and so on for any number.

So which is the result?
Let's leave this in the bunch of cold cases for now.
Let's instead look at how zero moved from India to Europe.

In 662 in Damascus, capital of the new Islamic empire that was emerging, was the Nestorian clergyman Severus Sabokt.

The Indus river is the longest and most important river in Pakistan. It is also the longest river in the South Asian continent, and third in the world for yearly flow.

Sabokt was the first one to report that there are refined mathematicians beyond the Indus, who utilise a numeric system which is superior to any other in practicality and efficacy, and which is founded on 9 digits plus zero.

Already in the 7th century then, Islam adopted the

Indian positional system and brings it over the Asian and African shores of the Mediterranean

This is the numeration that is referred to by the Arab mathematician al-Khwarizmi, author of the famous Al-jabr, which in Latin in translated as Algebra.

As for our research on zero, let's remember that the mathematician delineated the operation whereby, if one of the elements of an equation appears with a subtraction term, this may instead be added to the other element. In practice, before this the very notion of negative quantities was not considered.

Al-Khwarismi had the merit of illustrating the system of positional numbers from India based on only 10 digits.

Then, on both sides of the Indus, zero has gained the importance and concreteness of a number.

From the 7th century we see reports of numbers consisting of these 10 digits according to a positional system, where every symbol has a different meaning depending on the position it takes.

This system is also used today, and that is the reason why it is also called Indo-Arabic as well as positional.
But what is our dear zero called?

As we have mentioned, the Arabs call it *sifr*. The

Latins translate this as *cephirum*, meaning sapphire, known mythological figure personifying wind. From the year 1000, the Western world imports the decimal positional system from the Arabs, and in particular the existence of zero, which is contemplated from the 13th century.

Before the Indo-Arabic system was imported, the existence of zero was neglected, which entails a particular view of many things.

For example the year of Christ's birth is not year zero, but year 1, and in counting the years we go directly from the year 1 B.C.E to 1 AD

It is attested that Gerbert d'Aurillac, who then became Pope Silvester II, already used an abacus based on a rudimentary positional system in the year 1000.

Silvester II was born in Aurillac, France, in 950 and was the 139th Pope of the Catholic church, from 999 to his death. He is also the first French Pope.

Passionate about his studies, he introduced the Arabic knowledge of arithmetic and astronomy to Europe. In fact, Gerbert was the author of several treaties of mathematics and geometry. Despite his work, the decimal positional system did not take off right away in Europe. The turning point was with Fibonacci, who really lifted this system and caused it to spread.

Leonardo Pisano, known as Fibonacci, was born in Pisa in 1175. He is considered one of the greatest mathematicians of all time. He contributed to the rebirth of science after the decadence following the Middle Ages.

Leonardo was the son of a merchant, Guglielmo Bonacci, who traded with the Arabs. Towards the end of the 12th century, Guglielmo had a business in northern Africa. Thus Leonardo spends several years in Bugia, Alegria. Here he came into contact with Islamic culture and learns Arabic and becomes interested in mathematics. He was passionate to the point that he starts travelling to master his proficiency.

Reports tell us of his travels to Cyrenaica, a region in westerns Libya, to Egypt, Syria, Turkey, Greece and Provence.

Fibonacci became a great mathematician, capable of brilliant and innovative intuitions.

When he returned home to Pisa, in 1202, he publishes the text *Liber Abaci*. Thus he introduced positional numeration to Europe.

Before Fibonacci, the only mathematicians from Europe were Archimedes from Syracuse, in Hellenic times, and the Arabs from al-Andalus and Septimania, that was the part of the Iberian Peninsula governed by the Arabs.

Septimania is the ancient name for the French region of French region of Languedoc-Roussillon. The name dates back to the ancient Roman times, when the 7th Legion (in Latin Septima) was there, and lasted until the Middle Ages. Septimania covered about the same area as modern day Languedoc-Roussillon, apart from some parts of Gard and Lozere. Here, there were several mathematicians, scientists, and scholars in general, who continued the legacy of ancient knowledge, with fruitful outcomes, reaching new heights.

These medieval mathematicians were born in European territories, but belonged to cultures other than the Romans', while Fibonacci was the first great European mathematician. Thank to him zero began its ascent in Italy and Europe.

With him the Greek Euclidean geometry and mathematical tools developed by the Arabs and Alexandrians came together.

Fibonacci translated the term sifr into zephirus. From this we finally obtained zero. Even the term for cipher descends from sifr.

Such are the remarkable merits of Fibonacci, but unfortunately he was not updated on the latest developments in Indian maths.

In 1228 he wrote:

Novem figure indorum he sunt 9 8 7 6 5 4 3 2 1
Cum his itaque novem figuris, et cum hoc signo 0, quod arabice zephirum appellatur, scribitur quilibet numerus.

Which translates into:

For the Indians there are 9 digits: 9 8 7 6 5 4 3 2 1
With these 9 digits, plus the sign 0, which in Arabic is called zephirum, it is possible to write any kind of number.

It is clear that he studied al-Khwarismi's work, but also that he did not go far beyond this, despite his numerous travels looking for new ideas. So much so that he proposed a symbol for zero that is closer to the idea of nothingness than that of a real number.

Despite the innovation, the Indo-Arabic system took long to spread, maybe because it was not fully understood.

Moreover, to be honest, many mathematicians did not see the need for zero, since the four operations were carried out with an abacus or on paper, like we do today. In fact, in 1280, in Florence, this system is forbidden to bankers, since, they thought, the digits could be easily counterfeited and the system could be used to communicate some other kind of message. Because this system was called "cifra", we derive from it the phrase "ciphered message", i.e. a message written in code.

It was only in the 16th century that the Indo-Arabic system became thoroughly accepted in Western Europe.

So in the old world we got to know the wonders of zero relatively late.

Before us, it was known in Asia and Africa. Even pre-Columbian civilisations were familiar with the number zero.

Well then, let's get back to the talk about operations now that we understood how zero as a number was introduced.

The mathematician Mahavira, as we said, defined how zero is handled in the 4 operations, but he did not convince us about division, for which we expect some clarification.

To summarise, we have

Addition: $x + 0 = x$ and $0 + x = x$ (that is, 0 is a neutral element in relation to addition)

Subtraction: $x - 0 = x$ and $0 - x = -x$

Multiplication: $x \times 0 = 0$ and $0 \times x = 0$

Division: $0 / x = 0$ if x is not equal to 0.

So far so good.

But when 0 is the denominator, as in $x / 0$, we find that the expression has no result, since 0 has no inverse, as a result of the previous rule.

If zero is the exponent of a number, we have $x^0 = 1$, apart from the case where $x = 0$, which can be left undefined in some contexts.

Then there is 0 factorial:

$0! = 1$

Where the factorial of a natural number n, written n!, is the product of the positive numbers smaller or equal to n.

For example, 4 factorial is

$1 \times 2 \times 3 \times 4$

By convention of the empty product, $0! = 1$

So all the operations are convincing except $x / 0$.

To get some clarification, and possibly reassurance, on the operation $x / 0$ we must wait for the introduction of calculus.

21 CALCULUS

British mathematician John Wallis was born in Ashford, in Kent, in a rainy November in 1616. Between 1643 and 1689 he was chief cryptographer of the UK and of the royal court.

Wallis added and important element to the characterisation of zero. In fact, he contributed quite decisively to calculus, the field of mathematics that studies functions through the notions of continuity and limit. To him is owed the symbol ∞ denoting infinity.

So what little did not convince us earlier, may now have some sense now, thanks to this very notion that the operation $x/0$ takes us closer to infinity.

Indeed, if $0/x = 0$, we k now that $x/0$ has a result, and that is infinite.

In other words $x/0 = \infty$

Well then, we have now figured out zero as a number. A number which in the positional system is before 1.

But the do we count from 0 or 1???

Humans normally count from 1. We saw that from Roman times, we counted from the year 1 B.C.E. to 1 A.C. exactly because there was no notion of counting from something other than 1.

Today in informatics zero is a very popular starting point.

It is fundamental to differentiate zero from the letter O.

Some machines use a character for 0 with a bar crossing it. Another convention used for the first printers consisted of leaving 0 unchanged and adding a little serif to the letter O, making it look like an upside-down Q.

The character used in European licence plates distinguishes zero from O by drawing 0 more as an oval, while O is more round.

To avoid ambiguity, Germans also use a font called fälschungserschwerende Schrift (FE Schrift), which means *script complicating counterfeiting.*

The characters used in the UK do not differentiate zero and O because there should not be ambiguity if the correct spacing is respected.

In Italian licence plates, both old and new, the problem is solved by prohibiting the use of the letter O.

To avoid confusion, I, Q, and U are also not used.

In all cases where an error would be fatal, zero is avoided by all means.

For example, the booking codes used by Southwest Airlines only use the letter O and I instead of 0 and 1.

To summarise, the symbol for 0 is similar to an oval, while O is precisely round. This notion, which may now seem insignificant, will show us a new way of conceiving zero.

Another thought on zero: if it's a number, is it odd or even?

The answer is that zero is considered even. Because it comes before 1, and 1 is odd, zero is automatically even. But mathematicians have gone deeper.

First, we see that the set of natural numbers, i.e. the set of non-negative numbers $\{0,1,2,3,4,5...\}$

contains 0. Then, zero is before 1, which is odd, and thus zero is even.

Moreover, the set of even numbers is obtained by multiplying natural numbers by 2.

Therefore, since natural numbers are
0,1,2,3,4,5,6...

The set of even numbers is
0,2,4,6,8,10,12...

And the set of odd numbers is complementary to this
1,3,5,7,9,11....

22 MISTREATED ZERO

Let's review the concepts.

After learning how to count, man invented numbers and then acquired the notion of zero and represents it with a symbol that is unambiguous. Zero is oval, the letter O is round.

Multiplying zero by any number gives zero, but dividing any number by zero gives infinite, while dividing by infinite gives zero.

The relation between zero and infinity is clarified by calculus, which was developed in the 27th century thanks to Gottfried Wilhelm von Leibniz ad Isaac Newton, who both owe a lot to John Wallis.

Gottfried Wilhelm von Leibniz, latinised into Leibnitius, was born in Lipsia, Germany, in 1646.

He was really a brilliant man. Other than a

mathematician, he was also a philosopher, scientist, logician, glottologist, diplomatic, jurist, historian, and lawyer.

The term *function* is owed to him, and is used to refer to the properties of a curve, such as direction, slope, and chords.

Leibniz re-organised the mathematics of the 600s and is considered the precursor of informatics, neuro-informatics, and automated calculus. He invented the mechanical calculator, a.k.a. the Leibniz Machine.

Many aspects of his philosophy opened several paths into the dimension of sub-consciousness, which only in the 20th century exploded with Sigmund Freud.

The Englishman sir Isaac Newton, born in 1642, is known mostly for his contribution to classical mechanics, but also contributed to several more branches of knowledge.

In 1687 he published *Philosophiae Naturalis Principia Mathematica*, in which he described the universal law of gravity. He contributed to the scientific revolution and to the progress of the heliocentric theory.

Newton and Leibniz, together, are accredited the merit of introducing the first elements of calculus, in particular the notion of integrals, which is still used today with many of their notations.

Another development in the 28th century is owed to the introduction of limits, which in maths serves the purpose of describing the tendency of a function of approaching some value.

Thanks to the introduction of the notion of limit, the functions of derivative and integral are defined as limits and not as ratios or sums of infinitely small units.

There zero starts to become more concrete and to produce amazing results.

It is reintroduced, as we saw, in all fields of human knowledge, both as an abstract concept and as a mathematical number. In both cases it is rich, full on content and several nuances. A kaleidoscope that returns emotions to whom chooses to observe its contents.

Let's take Abstractism for example, the artistic movement that was born in the early 20th century, in parts of Europe far from one another and where it starts with no common intent.

Abstractism refers to those visual arts that avoid the objective representation of life.

Russian painter Wassily Kandinsky (1866-1944) is

considered the father of Abstractism.

Kandinsky had a precise conception of zero.

In his book *Point Line Surface* he writes:

The geometric point is and invisible entity. It must be defined as an immaterial entity. Materially, the point is equal to zero. But in this Zero several properties are hidden, which are human.

We represent this Zero – the geometric point – as associated with maximal concision, meaning with extreme reserve, which nonetheless speaks. In this way, in our representation, the geometric point is the highest and absolutely the only link with silence and speech.

It is therefore the geometric point that has found its material form, firstly, in writing. It belongs to language and signifies silence.

Here, if the point is geometric nothingness, a missing point is a black hole with zero dimensions. So we have a double image of nothing.

And we have two roads; one of conceptual zero and one of numerical zero, which meet here.

We must however introduce the difference between operator zero and medial zero. The former is the one that added to a number indicated that it assumes a value as many times as big, according to the base (10 in our system).

Adding a zero to 23 gives 230, that is a number 10

times bigger than 23.

In this sense, zero does not mean nothingness.
As for medial zero, this represents absence.

If we take 203, for example, here zero means the absence of tens (203 = 2 hundreds, 0 tens, 3 units). This is a fundamental difference between zero's positions.

If zero indicates absence, then the root of zero as nothingness is definitely in medial zero, not operator zero.

So at best we can understand this interpretation of zero as emptiness, or as lack of something, which is medial zero, and which is nonetheless different than the idea of absolute lack.

Here we mean lack of something.

But even operator zero is mistreated. As we saw, many consider it empty of value in itself.

According to the Italian thesaurus Petrocchi *Zero is by itself of no value, but to the right of digits it multiplies them by ten.*

In the Spanish thesaurus Molinier, the saying "left side zero" is mentioned (still used in Spain), which means something that has no value.

Even the French thesaurus Littre defines zero as

une chiffre qui de lui-même ne marque aucun nombre, mais qui, étant mis à la droite des autres, indique qu'ils prennent une valeur dix fois plus forte.

That is, a digit that has no value of itself, but that may indicate a value 10 times larger with other digits.

Here again we underline our dear poor little zero seen as nothing.

We mention one more case.

Giulio Raimondo Mazzarino was born in Pescina, Italy, in 1602. He became a cardinal, politician and diplomatic. He was active in France where he served King Luis 14, following cardinal Richlieu.

Between Mazzarino and Anna of Austria, queen of France, wife of Luis 13, a strong love story was born, and many believe he was the father of Luis 14, the sun king.

More or less at the same time, in 1648 to be precise, The Parliamentarian Frond took off in France, a movement revolting against Mazzarino from the Parisian Parliament and this also against Anna of Austria, who was reigning on behalf of her son Luis 14.

The Parliamentarian Frond was part of a larger movement at first, called the Frond, which opposed the absolutist tendencies of the French kings.

Even the bishop of Paris, Paul de Godi, who was better known as the cardinal of Retz, was an activist for the Parliamentarian Frond.

So the fight saw the support of clergy, nobility, and bourgeoisie, without anybody talking back until Armand of Bourbon comes along.

Which is why we started this discussion. Armand, prince of Conti, started the parallel branch of Bourbon-Conti, different from Bourbon-Conde.

The second male child of Enrich II of Bourbon-Conde and Charlotte Margaret of Montmorency was in fact the first prince of Bourbon-Conti, a title which was instituted especially for him.

His god-father was Cardinal Richelieu, and his god-mother the duchess of Montmorency. His brother was Luis II of Bourbon-Conde, known as Great Conde.

Prince Armand of Bourbon was not particularly brilliant, and Paul de Godi said of him that he *is a zero, that multiplies only because he is prince by blood.*

Even in the 600s and in revolting circles zero was used to refer to someone who is worth nothing. A crazy habit that spans social ranks and cultural levels.

With no solution of continuity, zero is brought up inconsiderately.

23 RABBINIC HEBRAISM AND KABBALAH

Let's now leave Parisian nobility behind and move onto the fascinating world of rabbinic Hebraism, and here, be prepared, the interpretations of zero might surprise us.

Kabbalah is the group of mystical and esoteric teaching of rabbinic Hebraism, spread already in the 12th to 13th century. The Tree of Life represents the universe in Kabbalah. Its description is considered as the cosmology of Kabbalah mystics.

At the top of the Tree of Life is a crown, which represents the point where divine influx is manifest from nothing.

One of the most important theorists of Kabbalah was Azriel ben Menahem.

Ben Menahem was born in Gerona, Catalonia, in 1160. He was a philosopher and mystic of Jewish faith. His teaching is different from that of other Jewish mystics due to some peculiarities. In particular, he refuted the theory of a creation from nothing.

According to Azriel, God's original will is to be interpreted as nothing. Creation originates from the movement that transforms infinity into emptiness, an abyss within God itself. Other Kabbalists did not want to accept this idea of coexistence between infinite and nothing, and conceived nothingness as a created entity.

Complicated? Sufficiently. But the conclusion of all these discussions is that however one gets to face the notion of zero, be it through mathematics, philosophy, art or any other path, this will divide people. But the dynamicity of zero, its movement, which is capable of generating other movements, movements of mobs and cultures, which may divide or unify, is part of its mystery and its appeal.

The only point which we feel we solidly solved, very convincingly, not as an act of faith but with certainty, is that zero is not nothing.

24 THE FULNESS OF ZERO

In the previous chapters, we dwelled on the notion of zero as nothing, nothingness, emptiness, interpreted and used in several scientific and cultural circumstances.

Now we are more serene. We have made our choice: zero is not nothing. And with this new consideration for zero, we embark again on this journey, this time interpreting it as universe, fullness of everything.

Let's start from Giotto. Yes, we want to bring up the great painter, sculptor, and architect, born in 1267, more or less, since we do not have reliable sources.

Giotto, also known as Giotto of Bondone, predates the so-called Renascence art which developed later in Florence, from 1400, and which is full of innovations, which spread out to several branches of art.

In visual arts, the intuitions of Italian artists like Giotto and Ambrogio Lorenzetti, or the French miniaturists, are studied with great rigour, all the way to producing revolutionary results which constitute the elements of the new style which from Florence spread to the rest of Italy and then Europe, until the early decades of the 26th century, when late Renaissance took place with Leonardo Da Vinci, Michelangelo, and Raffaello.

Giotto worked in a period when it was possible to free creativity from the limitations of the Middle Ages.

He was a visionary, but so was a Pope: Pope Boniface 8th.

Benedict Caetani was born in Anagni in Latium, in 1230. In 1294 he became the 193rd pope of the Catholic church with the name Boniface. Descending from an important branch of a Pisan family, he acquired more riches and great lands by using his pontifical role. He was strongly opposed to the coeval transition of European states from monarchies to national states, and he also staunchly contrasted political movements undermining the legislative power of the Church. He was one of the most controversial Popes of the Middle Ages, both in his time, and later.

So what brings together Giotto, Boniface, and zero?

For his jubilee in 1300, Boniface ran a sort of competition, to select a painter to commission a series of paintings.

A messenger was in charge of collecting the works and bringing them to the pope as an example of each competitor's work. Among the many artists of the time, one was Giotto, who was only 33, more or less, and who had not yet reached fame. The messenger asked him what he'd submit to show the pope his characteristic art style. According to Vasari, famous painter, architect, and art historian, Giotto had nothing to submit, but painted a perfect circle on the spot with only one stroke. Then Giotto told the messenger that the value of that painting would be acknowledged by the pope. Indeed, when the pope saw the circle, he understood that Giotto was superior to all other painters of the time.

This story became so famous that it is still told today.
But here the truth comes out!

Giorgio Vasari was born in Arezzo on July 30th 1511. His artistic formation was complex, based on early Mannerism, Michelangelo, Raffaello, and Veneto's culture. As an architect, he was a key figure of the initiative promoted by Cosimo I de' Medici, contributing to great building sites in Florence and Tuscany. Among these the Uffizi in Florence, and the renovation of Palazzo Vecchio, also in Florence.

Well, a great curriculum surely. But be mindful, we must look at the dates. Vasari was born at least 200 years after Giotto.

So, can we be sure of the story about the circle? Or even that it was interpreted correctly?

Vasari behaved like most art critics of the times before him and after. He had the presumption of knowing every piece of art not only through his, but also through the author's state of mind when he made it; knowing his secrets during the process of creation, which then ended up in his judgemental hands.

But Giotto actually drew a big oval, not a circle, in front of the pope's messenger!

Then, given what we've said so far, and knowing that zero is similar to an oval, we can claim that Giotto chose to draw an oval knowingly (representing a zero), and not a circle.

Maybe he intended to express the several, infinite sides a universe may contain? Probably!

Not a letter O, as many sustain, Vasari before all, but exactly an oval. Maybe to express the fullness of zero? Very, very likely!

Knowingly picking a symbol that represents Everything and not simply a letter.

First of all, an oval is much more difficult to draw free-handed. The concept it expresses is always that of zero, which is not Nothing, is it not nothingness. Zero is a set of complex ideas, concepts, and abilities.

With a single stroke, Giotto expressed not only his technical ability, but also the depth of his thoughts, the volume of his soul. The colours of his spirit.

And the pope saw all of this. He saw all of this universe in that sign. More than sending him a whole bunch of works and volumes to prove his personality.

In that sign, which is Everything, he saw the universe of Giotto, his marvel, mystery, and originality. He picked him.

There, we saw that zero is an oval and not a circle. Oval... like an egg. This little step brings us to a new world, a new way of conceiving zero.

25 THE PRIMORDIAL EGG

But what is zero, which looks like an egg? Can it be nothing (something of little value), if the same concept occurs in all ancient civilisations?

The *cosmic egg* is an expression of fecundity, it is the origin of all. It encompasses a profound meaning, as we will see, which seems to be with mankind from the beginning of time.

The *cosmic egg*, or the *egg of the world*, is a cosmological archetype.

It is the original model. Archetype comes from Greek *arche*, meaning origin, and *typos*, meaning model.

It is opposed to stereotype, from Greek stereos, meaning copy.

So the archetype is the first complete universal principle, of which stereotypes are a partial imitation.

It is different from prototype, which is simply the first element of a series of reproductions (based on an archetype).

The notion of cosmic egg is found among the Sumerians. Or rather, they already knew its conceptual structure in 2000 B.C.E.

The notion of cosmic egg in Mesopotamia spread to India. We find it in the Hindu religion.

In the 6th century B.C.E., Orphism spread through Greece, a big religious phenomenon of a mystic nature.

This century is important in the history of religions on a global level. It brought us a wave of illustrious people who somehow re-awakened consciences in lands far apart: Confucius and Lao-Tse in China, Buddha in India, Ezekiel among the Israelites, Zarathustra in Iran, and Pythagoras in Greece.

For the Orphists, the primordial egg is the origin of life, its fullness itself.

In Orphism, the cosmic egg is silver.

From this egg, laid by Night, ancient deity, and hatched by a blow of Northern wind, Eros is born.

Thanks to Orphism, the concept of primordial egg spreads to Egypt and Greece, until it reaches the

Pelasgians, an ancient people, forefathers of all Indo-European people.

Pelagianism was a Christian current. Pelagius mantained that the original sin did not taint human nature and the will of humans itself is capable of choosing Good, without the need for divine grace.

These theories were fought by Saint Augustine, and definitely condemned by the Catholic Church as heretic in the council of Ephesus in 431.

In the myth of the Pelasgians, the goddess Eurinom, having emerged from Chaos, lays the primordial egg.

The primordial egg's journey is strange and peculiar. Doesn't it remind of the journey of our zero?

The notion of primordial egg is present in oriental, western, and African religions, in China, and in Celtic religions in Europe.

In Hinduism, the cosmic egg is *Hiranyagarbha*. It is described in the books of *Bhagavadgītā*, or Song of the Divine. These books are made up of 700 verses divided in 18 volumes, and are part of the great epic *Mahābhārata*, in the version called *vulgata*.

Bhagavadgītā holds a sacred value, and for Hinduists it is among the most prestigious and loved texts.

The cosmic egg expresses the idea of an infinite universal nucleus in darkness, from which Lord Brahma makes it manifest.

The process is achieved through Aum, a syllable that allows respiratory emission, which in Hinduism represents the original breath of life.

From this creation the universe is developed, until its conclusion in maximal degrade. Then it starts over again in a series of cycles, called kalpa.

Mithraism is a religion of Greco-Roman origins, based on the cult of a God called Meithras, who only apparently derives from the Persian god Mithra and other Zoroastrian deities.
This God is often represented as appearing inside a golden egg.

In the Taoist religion in China, the cosmic egg is described in the myth of Pangu, the creator of the world.

For the Celts, this egg is called Glain.

For the Bambaras of Mali, in the beginning there was an empty egg that is filled and develops thanks to the creative breath of the spirit.

In the Polynesian myth Vari-Ma-Tetakere lives in a cosmic coconut.

In the ancient Egyptian religion, the Phoenix lays the egg, from which it is cyclically reborn. The Phoenix has a vital breath for which the God of air Shu is born. Nearing its death, the Phoenix builds a nest in the shape of an egg and there it burns, but from this combustion an egg is generated, which the sun hatches.

In Christian art, an egg in the hand of the Virgin Mary bears connotations that then are symbolically manifest in the festivity of Easter.

In alchemy, the egg of philosophers, is the symbol of the interior transformation, from raw material to philosophical gold. The so-called Great Work.
But this is not it.

Mircea Eliade was born in Bucharest in 1907. He was one of the major historians of religions, as well as a writer, orientalists, mythographer, essayist, and scholar. He spoke 8 languages, and read the holy texts in the original language. In his 79 years of life, he analysed many common traits of religions. According to him, the cosmic egg represents the repetition of the birth of the cosmos, the imitation of cosmology.

A concept such as the cosmic egg, that is oddly shared by all ancient civilisations, even those geographically distant from each other, and with no way of communicating.

The cosmic egg, from which all begins, is found in Polynesia, India, Indonesia, Iran, Greece, Phoenicia, Latvia, Estonia, Finland, Sweden, Russian, West Africa, Central America, and South America.

Even in more recent times a common origin for all things is sought.

Starting from the theory of relativity by Albert Einstein, with simplifying hypothesis where necessary, such as the homogeny and isotropy of the universe, in the 19th century big bang theory was affirmed, a cosmological model based on the idea that the universe began to expand at high speed in a finite time in the past, starting from a condition of extremely reduced volume and extremely high temperatures.

The process of expansion still continues now.

Here's something interesting with the progress of science.

From the 30s of the 19th century, astrophysicists began to talk about a primordial nucleus, pre-existent, unknown and unknowable. It is from this nucleus that the universe developed through the big bang, from which things are knowable, since they emit light.

In practice these scientists integrated the observation by Edwin Hubble, great American astronomer and astrophysicist.

Another important contribution to the theory that

everything shared a common primordial origin is made by Schrödinger.

Erwin Schrödinger was born in Vienna in 1887. Until his death in 1961, he made important contributions to physics, mathematics and quantum mechanics.

The Schrödinger equation is a fundamental equation that determines the temporal evolution of the state of a system, for example a particle, an atom, or a molecule. So important that it gets him the Nobel prize for Physics in 1933.

Schrödinger was passionate about Vedanta, part of the traditional Vedic literature, and defined his studies by taking inspiration from Indian cosmological philosophy.

According to modern cosmology, 3.7 billion years ago, or even earlier, the whole mass of the universe was compressed into a volume of about 3 times the size of our Sun. In practice we had some sort of primordial egg, a formed nucleus, or, according to Italian astrophysicist Vittorio Castellani, a broth of quarks, leptons, and photons. From this began the existence of the universe.

26 ZERO IN MANTIC ARTS

The mantic arts also support the idea that zero is full, a container with meaning, and not emptiness.

Since the beginning of time man has tried to find answers to the unknown. We tried to fight the unknown to gain advantage. Hence mantic art, that is the (presumed) ability to obtain information from supernatural sources inaccessible to most. The art of divination expresses itself through rituals, and may be based on the interpretation of signs, events, symbols, or presages, or even manifest itself through revelations.

Even today we try to interpret anything. No matter what, in good faith or otherwise, if it may grant us knowledge.

Then, if we celebrate zero, how not to remember its mantic value?

Let's consider, among the tools of divination, tarots. They are fundamentally a set of playing cards, in which zero is represented, normally associated with the fool or the jester.

Tarots were invented probably in Italy, in Ferrara, in the period between the end of the Middle Ages and the Renascence, even though in the late 700s Antoine Court de Gebelin promoted a thesis that held that the game traced back to the ancient Egyptians.

Court de Gebelin, from Nice, born in 1724, was a scholar of literature and an esoteric. He left the ministry as a protestant preacher to dedicate himself to studies, contemplating some of the esoteric trends of his time, from Rosa Croce to hermetism, from the theories of the Swedish mystic Emaniel Swedenborg, to the esoteric masonry, without neglecting Kabbalah. He authored a work in which he maintained that humanity lived in a lost golden age, before fragmenting itself into several civilisations and losing its knowledge. He was convinced that tarots derive from the will of the ancient Egyptian priests as tools to hide their wisdom. With this, he began the esoteric tradition that considers them a source of arcane knowledge.

Thus, thanks to Court de Gebelin, or partially thanks to him, tarots, that were initially just a game, began to be used for divination in France in the 28th century.

Different tarot decks of cards were created throughout the centuries. Frequently, for example, noblemen would commission their own decks to represent their houses.

Though having different artistic characteristics, the cards always shared the basic principles.

They are always 78 cards. The group of the major arcane (which is more meaningful from a mantic standpoint) is made up of 22 cards, illustrated by human, animal, or mythological figures. The group of minor arcane is made up of 56 cards, divided into 4 series of suits in the Italian tradition: cups, coins, swords, and clubs.

Here we get to zero.

The major arcane are numbered from 1 to 21. Then there is the Fool, representing the arcane zero, which has a special role.

The Fool is represented dressed in rags. In some decks he is barefoot, in other decks he wears socks and bears a cane on his shoulder with a bundle tied to it. What is in the bundle? His whole world. His material world, sure, but also his thoughts, his dreams, and ambitions.

The Visconti-Sforza decks date back to the 25th century, and originated the classical decks, in particular the Marseille variant from which the modern decks derive. They are of historical and artistic interest, for the beauty of their illustrations, realised with precious materials, and in some cases, portraying members of the families Visconti and Sforza. According to Stuart Kaplan, scholar and collector of tarots, there are 15 decks belonging to the Visconti-Sforza group.

In these decks, the Fool of the arcane Zero is represented with some feathers in his hair. Some think this was inspired by the Stultitia by Giotto, a fresco from 1306 in the Chapel of the Scrovegni in Padua, Italy. Stoltitia shows the profile of a male figure. He is dressed as a jester, with feathers in his hair, a skirt of rags, a braid around the waist from which sphere are hanging. On the other side it pairs with Prudence, virtue of who ponders on their choices.

In the Marseille tarots, the Fool, our zero, is represented as a jester. This representation is also adopted in other decks and surely influenced the representation of the Jester in modern cards, which in Italian card games has the same role as the fool. So it is an important card, and not one to be discarded.

By the end of the 800s, Aleister Crowley did not abstain from realising his own deck of tarots, which had obscure aspects, though containing the jester.

Crowley was born in 1875 in Leamington Spa, in England. He was a controversial esoteric, artist, writer and alpinist. Many consider him the founder of modern occultism, other think of him as the inspiration for Satanism. He is also considered a key figure in the story of the new magical movements. Crowley is credited with the most relevant attempt to create a magical religion for his time. His influence on the magical sphere is fundamental.

27 ZERO EMBRACES TRASCENDENCE

In Crowley's tarots, the Fool, i.e. Zero, is represented by the Greek god Dionysus. This deity represents the deep nature of mankind, our primordial, savage, instinctive part, that is present even in civilisation. As an implacable original part, it may emerge and explode violently if repressed, instead of being understood and channelled correctly.

Dionysus is often represented on a chariot with his partner Arianna. His priestesses are the Maenads, or Bacchants, women taken by frenzy and possessed by the deity. As a deity of vital force, impulse, and drunkenness, Dionysus was often the subject of the analyses of Friedrich Nietzsche, who juxtaposed the Dionysian spirit and the Apollonian one (from Apollo), indicating reason and balance.

Dionysus for the Romans is Bacchus. Now, if we

may associate Bacchus with drunkenness and the inhibition of the senses, it is also true that the Romans claimed "In vino veritas" meaning "In wine is truth".

The meaning of this saying is that when one is under the influence of wine, of alcohol, they have no inhibitions and may more easily reveal facts and thoughts that they would not share sober.

So again, the Fool, the arcane Zero, is the most interesting, granting revelations with no filters.

Let's summarise the meaning of mantic zero. While this card represents foolishness, something undesirable, it later becomes madness, and the card has the same value as the King of Coins and the jester. Therefore an important one. The most desired, which implicates the most possibilities.

As an esoteric symbol, foolishness is what allows to look at life anew and search from the beginning. This notion starts from the fullness of zero. We have learned that 0 has the meaning of universal multiplier, as a number. Each number multiplied by zero is still zero, it represents Everything, unity. Moreover, as we have seen, it is the first among all numbers, and represents a new beginning.

In mantic arts, in tarots, the meaning of the fool take on new features between innocence and madness, including wilderness, instinct, originality, thoughtlessness, and detachment from the mind.

In substance this represents the innocent and primitive part of man, which can lead us to Good as well as Evil. The fool, when represented as a vagabond, symbolises the search for change, the path to evolution. In a spiritual sense, he can also represent the passage to a much higher level of awareness.

In the mantic arts, zero is also the life force, which embodies the whole path and projects us towards a new beginning. It may represent novelty, often unpredictability, and unexpectedness.

In astrology it corresponds to the circle of the Zodiac, which represents a complete cycle, in 12 steps, which repeats spirally. During the first three string signs (Aries, Taurus, Gemini) there is a blossoming of life, the expansion of light, which strengthens the affirmation of the following summer signs (Cancer, Leo, Virgo).

These first 6 signs belong to the objective world. The autumnal signs (Libra, Scorpio, and Sagittarius) express experience with otherness, maturation through broadening of one's horizons, which allows the consolidation of what was undertaken and the possibility of projecting oneself towards the future, thanks to the following winter signs (Capricorn, Aquarius, Pisces).

Zero corresponds to the element of Air, and therefore oxygen, which allows life.

The signs of air, that is Gemini, Libra, and Aquarius, are characteristically dynamic, positive, extroverted and rational. People with these signs have similar features but express their social nature differently.

We saw that Kabbalah is the set of esoteric and mystical teachings of rabbinic Judaism, already present in the 12th-13th century. More broadly, the term refers to those movement born in Jewish circles around the end of the Second Temple. Even Kabbalah features zero.

Eliphas Levi, pseudonym for Alphonse Louis Constant, was born in Paris in 1810. He was the most famous occultist and scholar of esoterism of his century. In his esoteric school, zero corresponds to the letter shin of the Hebrew alphabet, representing divine power. Shin is one of the most important letters. In practice it represents God's two names: Shedai (enlightened) and Shalom (peace).

The 800s saw the flourishing of several esoteric schools. Samuel Liddell MacGregor Mathers was born in London in 1854. He was one of the most influential figures in the field of modern occultism, and mostly known as the founder of Golden Dawn, the hermetic order of ceremonial magic, of which several branches exist even today.

In the school founded by Lidell Mathers, the fool is very important and corresponds to the letter aleph, the first letter of the Hebrew alphabet (alpha in Greek). Then, again, that from which all began.

In the tree of life, which in Kabbalah represents the laws of the universe, our zero represents the crown, the centre of creative will, the inspiration of the universe.

Basically, all but nothingness!

The fool, as explained by Oswald Wirth, Swiss esoteric writer of the 800s, is "the salt that generates other salts, the immaterial substratum of all materiality, the fire of intellectual life".

So we moved from zero to foolishness and foolishness is here meant as immaterialness, but also as knowledge.

From here it is easy to move onto Shakespeare.

William Shakespeare was born in Stratford-upon-Avon, in the United Kingdom, on the 23th of April 1564. He is considered the most important writer of the English language and generally the most prominent play-writer of Western culture. Of his works we have 37 plays, 154 sonnets, and another series of poems. His texts are translated in all the major languages.

In his texts guess who plays a significant and fundamental role?

The fool, clearly! Let's not forget that until now we have followed a path joining the fool to our beloved zero.

Clearly, we can here make a distinction between the fool and the crazy.

While with the term crazy we tend to refer to a person affected with a psychological pathology, with the term fool we refers mostly to someone out of line, someone who surprises us and defies our canons.

Shakespeare gifts us the notion of truth expressed through sane foolishness.

The fool, is in charge of communicating the most profound messages of Shakespeare's plays.

The fool is granted immunity in a way. He may say the truth without paying the consequences.

The centre around which the character play is always revealed by the most humble character. The fool.

The fool is lucid in his madness and denounces real illusions, rejects false truths, laughs seriously.

This is what Shakespeare tells us. Aware of his part, the fool does not expect to be considered. But here also lies his strength.

He does not worry about judgement. He's the fool, guardian of all truths. It is up to the free will of the interlocutor to grant him credibility or not. The fool is free, raw. He contains infinity. Just like Michelangelo's marble block, remember?

The fool is constantly present in Shakespeare's works. Then, we cannot think of Shakespeare without thinking of the fool. Therefore, given this analysis, we can't think of the fool without his truths and his lucid madness.

Elizabethan theatre represents one of the artistic period of greatest splendour for British theatre. It developed between 1558 and 1625, during the reign of the monarchs Elizabeth I and James I. The term refers to the theatre flourishing in the period from the reformation to the shutdown of theatres in 1642, upon the civil war. A long period, that also included the reign of Charles I, successor of James I.

Shakespeare is considered the emblem of Elizabethan theatre, in which life is lived as a play within the stage that is the world. Here the fool and what he represents are at ease.

As you like it is a pastoral comedy in 5 acts, written by Shakespeare between 1599 and the early months of 1600. The author follows the events around heroin Rosalind while she escapes the persecution of her uncle's court and falls in love in the Woods of

Arden. In the play, there is one of the most famous soliloquies, *The whole world is a stage*, and the sentence *Why then, can one desire too much of a good thing?*

In this play, the fool affirms the will to be mad as the one met in the forest who communicates the truth which was until then unconceivable for his mind. Once again the fool is a guardian of truths.

We are not properly sure that Shakespeare wrote these, in fact, we're not even sure of his existence, or rather that the pieces attributed to him are really his.

The scarcity of documents about his private life is still cause of debate regarding this attribution.

The speculation consists in the controversy, started in the 18th century, on the possibility that the plays may be the product of another author or a group of scholars.

Anyway, regardless of who wrote the plays, the author seems to contradict himself, or maybe just to be a little confused.

Between 1605 and 1606 Shakespeare wrote *King Lear*, a tragedy in 5 acts, in verse and prose. The story is rooted in British mythology. It is a drama with two plots, in which the secondary one contributes to highlighting and commenting on the primary one.

In this work, the exchanges between the king and the jester show how important this figure is.

Can't you make any use of nothing, uncle? The fool asks the king.

Surely not, young man, nothing can be done with nothing. Answers the king.

But there is confusion.

You evened your brain from both sides and left nothing in the middle. Then: Now you're a zero with no digits. I am better than you: I am a fool, you are nothing.

When he says "zero with no digits" he means that zero with no digits to its left as we saw earlier, in other words the one that alone is worth nothing.

But at the same time he differentiates himself (the fool, the crazy, zero) from nothing.

So he claims:

I am the fool, you are nothing.

This is correct, since the fool, representing zero, it all but nothing. It is less correct to present zero on the right in the same sentence.

Anyway, apart from this little confusion (which is shared by many fellow writers, philosophers, and mathematicians), Shakespeare, giving power to the lucid madness of his fool, grants him a much more complex value than the mentioned zero intended as nothing.

Therefore, the fool is freedom. Firstly freedom of

expression. Freedom to express his own truth without fearing the consequences. But the zero-fool in Shakespeare is no hero, he is well aware of not running risks, he knows he is licensed and untouchable. Shakespeare promotes democracy in the most desirable way, without political, social or economic conditioning.

All of this embodied in the fool. All of this and much more in zero.

Thus Shakespeare is detached from the common sense of nothing-zero and the notion of zero-operator.

His instinct gives birth to a new approach to zero as universe.

28 THE EMANCIPATION OF ZERO

The notion of zero is surely hard to digest, when even the expression "being a zero" implies difficult connotations, before its meaning may finally be felicitously considered positive.

Being a zero is a great thing. But to be aware of this, not of being one, but of the value of one, and thus of the fantastic, marvellous responsibility of identifying with zero, requires effort.

We saw that a great number of mathematicians and physicists have tackled the issue of zero as nothing and as an empty sum, if not in relation with other digits. That is, they interpreted zero as a sign that has value only based on the presence of another number in a positional system.

Even more fascinating is the approach of the philosophers, many of whom are a mystery to most.

Some interpret it as nothing, some other as the negation of being, which then plays a nihilistic role, as if zero was equal to nothing and opposed to being, and negated it.

But this isn't so. Even if we wanted to refer to the Babylonians and conceive zero as absence (of something that was and no longer is), zero by itself is not negation of anything. But be mindful, this does not mean that it has no consistency and that it is, so to speak, inactive, meek.

The great philosopher Georg Cantor, who we saw was the father of cardinality, posited a theory of sets that contains infinities, that is many infinites, one within the other, as in a game of Chinese boxes.

Cantor summons God as the infinite of infinites, but also as the ultimate horizon where reason strands.

Be it as it may, the issue of God in relation of what we are discussing, zero, appeared open and now it no longer is. Solved, accounted for, on a physical-mathematical level before any other level, for example ethical.

Let's see why.

The Bible says: God created the world from Nothing.

How to interpret this?

God drew everything that exists apart from him from Nothing.

This notion finds expression in the first page of the scriptures, even though it is only made explicit in the following revelation of Genesis.

The statement "God created the world from Nothing" can be interpreted in only one way, if one wishes to avoid absurdity, or worse, madness. And by madness we mean the madness of a psychiatric patient, not the knowledge of the fool.

Some think that this statement in the Bible is not about a specific fact, nor the totality of facts, nor being, but simply the sense of being.

But this is the archetypical question: is life meaningful? And if it was, what is its meaning? And still: is there a meaning for life in itself, or is the one for each being creature, in other words a meaning for each individual's life?

When we say that God created the world from nothing, we mean that life in general, and ours specifically, has a meaning. Otherwise God would have left things in the nothingness from which he drew them: each thing that may be, living or not.

From the planets to the stars, form plants to mankind.

He drew all from nothing.

And this tells us two things. First, everything that is has a reason to be. Then, it tells us that nothing is really everything. It is what contains everything, before it may even be. Here, elsewhere, yesterday, tomorrow, and forever.

Let's remember Michelangelo. His marble block is like a small world. Within it there is already anything before it may be, before the master finally extracts it from raw marble.

Then to the astrophysicists, who try to explain how the universe was born and how planets and stars were formed, and to the philosophers and theologians, who instead seek a reason until exhaustion: why do we live?

To all who search, who ask questions, we can say with a single wonderful summarisation: zero. In zero are all answers.

Whoever seeks a reason seeks a creator.

But it may be that the reason is not there, not according to our knowledge, our attitudes, our culture.

Still, it may be that a creator exists.

Physics and the Creator and the Reason of the existence of things (everything, from stones to living creatures) may not be tied together in a bundle.In other words, if there is no creator, there still may be things that exist or have a reason. Likewise, if we have a creator, there may be no reason we can understand in what we see, or perceive, as existing.

Zero is Being in the absolute sense. Any "o" for some (for whoever wishes to adhere to the hypothesis here proposed) becomes an "is". It is. Like the verb to be.

With that oval, which was not an "o" as in the letter "O", but rather, unequivocally (we explained this earlier) a zero, Giotto expressed a nothing that is not nothing, he tells a story with a zero which contains absoluteness, his whole mastery, his knowledge, his intelligence.

Did Pope Boniface 8th choose that sign, that graphic symbol because he was fascinated by the perfection of nothing? Surely not, it is much more likely that he chose it because he was enchanted by the fullness of Everything. A notion as large as infinity, but concrete, tangible, like the world. An oxymoron of being, or rather its magnificent, manifest presence?

The fact is that thanks to this zero Giotto gains the commission. Surely not simply thanks to his dexterity in drawing an oval, which with a little practice even a rookie could achieve.
Zero cannot be nothing.

It is not what is not, but absolutely what is. It is not a *less*, but a *more*.

It's not the negation of existence, its subtraction, but instead a summation. The sum of existences, consistencies, that are summed two by two, not cancelling each other out, but giving life to another infinity of existences and consistencies, that in turn are summed, and so on.

To make it simpler, let's think of the universe of numbers.

Let's think of 3, a beautiful number.

+3 is a number. -3 is a number. If we add -3 and +3, we obtain a number. This number is zero.

Zero in maths is then a sum of numbers. We cannot imagine zero as not existing.

Then we can say that zero exists and has a double consistence compared to any other number. Since any number has an absolute equivalent, but of opposite sign, if we sum a number and its opposite we obtain zero. But if we sum 1 and its opposite, and then 2 and -2, and so on infinitely, we still get zero, which is then the sum of all numbers.

Oh man, can we now see the power of zero? Of how big it is? Then zero is the greatest number we can imagine. A concrete concept, like a number, but big, like a sum of infinity.

Our minds incinerate themselves even trying to grasp its greatness.

Then, if in order to concretely express this concept, whose comprehension, even an attempt at comprehension, makes neurons jump like corn in an oven, we can turn to something else, equally great and inexplicable with our limited means: God.

God is fullness, he is what contains everything. An infinite sum of finite sets.

And when we read "he drew the world from nothing" it means that he drew the universe from that everything that he himself is.
There are no contradictions in God and no annihilation.

Plus and minus have the same consistency, one allows the other to exist, mutually.

Let's remember the symbol of the Pisces in the Zodiac: two fish tied to one rod. One proceeds if the other recedes. It is the sacrifice of love, the gifting of oneself to one another. The symbol was adopted by most early Christians as a recognizable sign of identification. A coincidence? Maybe.
Let's take this notion to numbers.

Negative and positive are just adjectives that we attributed to these to comprehend them better.

No connotation of good and evil. Simply, the numbers with opposing signs can be joined, they can be unified in zero. And this is just for whole numbers. But among them are fractions.

A thousand universes, where the mind may not reach concretisation of neither connotations nor adjectives.

Then it is evident: if one is a number, aiming to be zero is the top.

Have you ever felt like a zero? Or rather, have you ever considered other to be like zeroes? We can only rejoice in this.

Whoever is called a zero is after all receiving the ultimate compliment.

Every zero is a line of fire, with a concentration of substance of unspeakable, stupefying beauty.

Not everyone can be a zero. Only those lucky enough, who feel the honour (or even burden) and the responsibility of defending their own infinite magical circle (by circle we mean a closed line).
Be he the lord of his reign. May he decide who shall be outside and who may be granted access (with

parsimony).

Not because of power, but rather because the circle, the boundary, is his own, and no one else's. Every circle (which to be precise, is an oval here) is unique, exclusive, personal. Those who have one can use it as they will. Clearly not everybody has one, meaning not everybody can be a zero.

To be a zero one has to pass a strict selection, where everything originates and where the infinite infinities are summed and everything is before it is.

Sometimes one is a zero without even knowing so, other times one wishes to be one.

Those who are chosen have a precious gift, like being granted access to the Akashi archives, the knowledge of knowledge, since all is in zero.

Every zero-world is perfect and beautiful. Full of wonder, full of intelligence.

Within this it is possible to speak to God (for those who contemplate his existence not because of faith, but rather certainty), but even with the preferred philosophers, and then with saints and heroes, not only those of history, but also the everyday ones, those next door. The lord of the circle has the authority to decide who enters: past, present, and future people.

Entering a zero-world, unless one you are its origin, is extremely hard and it is useless to try to find the rules, or looking for signs or belonging or a dress code, social or cultural status.

It is a decision for the lord of the zero-world to allow someone in or not. Without rules nor reasons, by virtue not of dictatorship, but a right, that which cost Eve and Adam the comfort of Eden, and made them (potentially) the children with the image of God: their free will.

Those who feel like atheists lose no hope, there are realities that go beyond faith, form and colour.

And zero, as we have seen, is intrinsic in the history of man. Beyond religion, culture, or place.

So long live zero. Long live all the special and lucky people who are zeroes.

To be a zero is indeed a wonderful privilege. For many, not for all.

To all the zeroes in the world, we ask here to get in contact with us. To share an intelligent exchange of opinions on a mystery of infinite beauty.

29 AFTERWORD

WHY FREE ZERO

Antonia is here with me.

Riding my car, life seems normal. After all this is the place where I feel most safe.

My friend Antonia is fundamental when I must reach a destination from my car. Be it a few metres or some steps, the journey is epic. Her embrace is a fundamental support to me. I have tried before to use crotches. But my primary problem is balance, and trying to find a centre of gravity for me and two other points is even harder.

Therefore Antonia is my optimal solution. Plus there's the pleasure of having her with me.15:50. I am in one of the most important and qualified hospitals in Milan.

I have checked in in more or less 40 minutes. A short time, for such a big structure.

I get to the 4th floor, where there are studios for visits to private doctors. In practice a patient can save some money compared to an actual private clinic, since the doctor is using a public structure to operate (which makes up part of the cost).

I look for the name of the practitioner who will take care of me.

Antonia and I sit on the anonymous black seats in the hall that works as a waiting room. We choose two seats right next to this practitioner's office.

It must be him.

Apparent age: 37. Tan. But judging by the time of the year – October – and by the amount of moisturiser on his face and neck, I suppose it is not a natural tan, but rather the product of a tan paid for in some beauty centre, possibly part of the hospital to which practitioners have privileged access.

His gaze, walk and look are those of who feels above the masses of common mortals. These clues bring no good prospects.

I am rarely wrong about people.

I hear my name.

It is suddenly my turn. Maybe I am the first patient on the list for this medic, who only works two afternoons in a week in this hospital, from 16:00 to 18:00.

I step up to meet this narcissistic doctor, who in seven minutes wraps up the visit, even finding time to refusing to look at the two folders of medical documents of all sorts that I had brought with me. He releases me saying

I have nothing. For my unbearable pain in my ear and nape, he suggests a buy an over-the-counter spray for inflammations.

We are talking about a famous doctor whose time costs me 150 euros.

The doctor doesn't even give me time to speak back and leaves the room without closing the door.

I see him walk through the hall and into a space which probably separates the patients from the medical staff: a lounge, toilets etc.

After ten seconds, enough time to follow him with my gaze up to the partition.

A few seconds, but enough to take a picture among the worst ever taken of doctors in the past 3 years.

And even this time my perceptions were right. am filled with anger.

I take my folders and wait again in the hall, ready

to attack. He must come back out from here if he has another patient.

My only worry is that I might be the only patient for that evening, or that there may be another way out. Another way to avoid my objections. I wait, if anything to let the anger out.

Antonia looks at me speechless. She knows my torment, the frustration I feel not being heard, believed. She knows my condition, which is worse day by day.

She looks and says nothing – for me this is perfect.

I see the doctor crossing the door again, always fast.

I stop him, I put the documents to his face, but after pretending to look at them, he claims he sees nothing: he can't rely on the "visionary" diagnoses of other medics.

He tells me that the source of my problems may be psychosomatic, due to stress, or who know what else… Hinting at some psychological frailty.

I try to use my last moments, I think he owes me this, given that I paid for his professional help more than 20 euros per minute, basically 1200 euros per hour. So I hold him with a visible gesture.

I throw the documents on the seats (in reality I would have thrown them at him if I was not afraid of damaging them) and I tell him with a decisive tone, while maintaining a low voice, that he probably got his degree by distance. A mistake, sure. While I said that I was thinking of the school model RadioElettra of Turin, the first example of distance learning, back in the 50s, when one would correspond with the professors by mail. Since we are in the digital era, and given the young age of the doctor, who surely he didn't know of the great Turin school, I should have mentioned some online school, maybe in Eastern Europe. Unforgivable mistake.

Anyway the allusion to an unsuitable formation, if anything regarding the approach to patients, which surely would have made Hippocrates turn in his grave, was grasped. He got my opinion. He looks at me disgusted, even more convinced of an emotional problem in me, and dashes to his office, closing the door behind him, if anything to highlight the insurmountable barrier between me and him. Annoyed, but not even much.

I am not hurt by his arrogance. Or rather, I am hurt, especially after spending years chasing different doctors, spending hours online looking for a reason for my symptoms, looking for a link between the things I felt in my body, considering it a single complex system in which everything is connected and not a bunch of compartments on which doctors, specialised in centimetre square large zones, could not work on without looking at the whole.

Yes, because with today's super-specialisations, an orthopaedist who looks at your hand knows nothing about your ankle, for example.

Anyway what hurts me primarily, now and many times before, is being considered less than nothing, just a tick sneezing, nothing more than a zero, with no content and treated as such.

Antonia helps me to gather my belongings,

I just want to get out of there.

I leave, sure, under my Antonia's arm, which sustains me, in all the senses. I leave with the usual load of bitterness, but with a certainty, a new one.

Today my reaction is not in vain.

What did I get from this little doctor?
Basically nothing. But after the 100th arrogant treatment of a specialist I decided I can no longer be quiet.

I made a decision. I will write something. Not on my illness, which by the way is called Arnold Chiari I, according to a diagnosis I received a few months afterwards, but about the years and years of switching doctors, as well as medical specialties, sanitary structures and so on.

It is a meagre satisfaction to know I was right to look at links between symptoms that nobody was connecting.

Now what I want is that nobody shall feel the punch in the stomach one feels when an arrogant smug person makes you feel like a zero, or at least try.

In my situation, but also in many other, even the simple everyday ones, it is necessary to have the courage to externalise one's own rebuttals, if possible. Without thinking that one may be right a priori, because of a tag on their gown, or a degree on a wall.

One must say their piece, always. An intelligent person listens. The exchange enriches knowledge.

My medical path is a long one, a very long one, and I am still walking. I have discovered many things, meaning that for me they were discoveries, and I will have no hesitation in sharing them with those who are ill, curious or just members of that weird species that are doctors.

But here, to the benefit of I hope many, I really want to simply make the beauty of the zero-universe clearer.

Nobody be afflicted when arrogant bullies, of any

age, take the right make another human being feel without consistence.

Zero is a magical universe, full of richness.

Truly, I just want to impart the dignity that zero deserves. The rest is simply a corollary.

So, the wish to free zero from unfair judgement was born in a hospital in Milan, after meeting a doctor unworthy of such qualifications.

I really wish to redeem it, to grant it the right acknowledgements. And I would be happy to share my perceptions with other zeroes.

Myzeroworld on

30 BIBLIOGRAPHY AND WEB LINKS

FOR THE ORIGINAL VERSION
Bottazzini U., Numeri (formato kindle), Società editrice il Mulino, 2015
Capello A. – Ferrari M, Numeri. Aspetti storici, linguistici e teorici dei sistemi di numerazione, Decibel, 1990
Crivelli Nadav E., I numeri del segreto. Manuale di ghematria e numerologia cabalistica, Psiche, 2011
Odifreddi P., Sorpresa, la matematica non è una nostra invenzione, Repubblica, 21 luglio 2000
Platone, Sofista, Armando Editore, 2006
Wassily Kandinsky W., Punto, linea, superficie, Adelphi, 1968
Zavattini C., Io sono il diavolo, Bompiani, 2003
Malika Lakon-Tay, Tarocchi & Cartomanzia, Edizioni R.E.I., 2015

https://it.wikipedia.org/wiki/Cardinalità
https://it.wikipedia.org/wiki/Georg_Cantor
https://it.wikipedia.org/wiki/Trilussa

https://italiangems.wordpress.com/2014/11/23/tales-
of-trasteveres-poets-trilussa

http://www.iltempo.it/cultura-
spettacoli/2012/08/25/quei-versi-di-satira-romanesca-
che-il-regime-rispettava-1.11724

https://it.wikipedia.org/wiki/Trilussa

https://it.wikipedia.org/wiki/Galileo_Galilei

http://cronologia.leonardo.it/mondo42f.htm

http://cronologia.leonardo.it/mondo42f.htm

https://it.wikipedia.org/wiki/Teoria_degli_insiemi

https://it.wikipedia.org/wiki/Ernst_Zermelo

https://it.wikipedia.org/wiki/Adolf_Abraham_Halevi_Fr
aenkel

http://webmath2.unito.it/paginepersonali/negro/ist/mis
ura.pdf

https://it.wikipedia.org/wiki/Misura_(matematica)

https://it.wikipedia.org/wiki/Nicolas_Bourbaki

https://it.wikipedia.org/wiki/Simone_Weil

https://it.wikipedia.org/wiki/Linguaggio_macchina

https://it.wikipedia.org/wiki/Codice_binario

http://www.tecnocino.it/2012/06/articolo/apollo-11-i-
computer-che-portarono-l-uomo-sulla-luna/39383/

https://it.wikipedia.org/wiki/Sulla_natura_(Parmenide)

https://it.wikipedia.org/wiki/Dike

https://it.wikipedia.org/wiki/Cartesio

https://it.wikipedia.org/wiki/William_Shakespeare

https://it.wikipedia.org/wiki/Amleto

https://it.wikipedia.org/wiki/Platone

https://it.wikipedia.org/wiki/Sofista_(dialogo)

https://it.wikipedia.org/wiki/Leucippo_(filosofo)

https://it.wikipedia.org/wiki/Democrito

http://materialismo-
atomismo.exactpages.com/leucippus.htm

https://it.wikipedia.org/wiki/Stagira

https://it.wikipedia.org/wiki/Arthur_Schopenhauer
https://it.wikipedia.org/wiki/Friedrich_Nietzsche
http://www.leopardi.it/
https://it.wikipedia.org/wiki/Giacomo_Leopardi
https://it.wikipedia.org/wiki/Buddhismo
https://it.wikipedia.org/wiki/Karl_Jaspers
https://it.wikipedia.org/wiki/Martin_Heidegger
http://www.eniscuola.net/2014/01/08/la-crisi-della-fisica-classica/
https://it.wikipedia.org/wiki/Max_Planck
https://it.wikipedia.org/wiki/Metodologia_di_misura
http://www.multiversoweb.it/rivista/n-11-misura/la-misura-il-problema-irrisolto-della-meccanica-quantistica-3505/
https://it.wikipedia.org/wiki/Fisica
https://it.wikipedia.org/wiki/Stephen_Hawking
https://it.wikipedia.org/wiki/Piergiorgio_Odifreddi
https://it.wikipedia.org/wiki/John_Cage
https://it.wikipedia.org/wiki/Basilide_di_Alessandria
https://it.wikipedia.org/wiki/Anders_Celsius
https://it.wikipedia.org/wiki/Muhammad_ibn_Musa_al-Khwarizm
https://it.wikipedia.org/wiki/Plutarco
https://it.wikipedia.org/wiki/Talete
https://it.wikipedia.org/wiki/Pitagora
https://it.wikipedia.org/wiki/Logica_matematica
https://it.wikipedia.org/wiki/Sistema_di_numerazione_posizionale
https://it.wikipedia.org/wiki/Brahmagupta
https://it.wikipedia.org/wiki/Mahavira
https://it.wikipedia.org/wiki/Indo
al-Khwārizmī
https://it.wikipedia.org/wiki/Muhammad_ibn_Musa_al-

Khwarizmi
http://www.treccani.it/enciclopedia/algebra_(Enciclopedia-Italiana)/
https://it.wikipedia.org/wiki/Papa_Silvestro_II
https://it.wikipedia.org/wiki/Leonardo_Fibonacci
https://it.wikipedia.org/wiki/Liber_abbaci
https://it.wikipedia.org/wiki/Calendario_maya
https://it.wikipedia.org/wiki/Matematica_babilonese
http://www-history.mcs.st-and.ac.uk/HistTopics/Egyptian_numerals.html
https://it.wikipedia.org/wiki/Cirenaica
https://it.wikipedia.org/wiki/Settimania
https://it.wikipedia.org/wiki/John_Wallis
https://it.wikipedia.org/wiki/Gottfried_Wilhelm_von_Leibniz
https://it.wikipedia.org/wiki/Isaac_Newton
https://it.wikipedia.org/wiki/Astrattismo
https://it.wikipedia.org/wiki/Fronda_parlamentare
https://it.wikipedia.org/wiki/Giulio_Mazzarino
https://it.wikipedia.org/wiki/Armando_di_Borbone-Conti
https://it.wikipedia.org/wiki/Albero_della_vita_(cabala)
https://it.wikipedia.org/wiki/Azriel
https://it.wikipedia.org/wiki/Arte_del_Rinascimento
https://it.wikipedia.org/wiki/Papa_Bonifacio_VIII
https://it.wikipedia.org/wiki/Giorgio_Vasari
https://it.wikipedia.org/wiki/Divinazione
https://it.wikipedia.org/wiki/Mazzi_Visconti-Sforza
https://it.wikipedia.org/wiki/Stoltezza_(Giotto)
https://it.wikipedia.org/wiki/Aleister_Crowley
https://it.wikipedia.org/wiki/Dioniso
https://it.wikipedia.org/wiki/Il_Matto
http://www.panouden.com/astrologia/cerchio_zodiaco.htm

https://it.wikipedia.org/wiki/Cabala_ebraica

https://it.wikipedia.org/wiki/Samuel_Liddell_MacGrego
r_Mathers
https://it.wikipedia.org/wiki/Albero_della_vita_(cabala)
https://it.wikipedia.org/wiki/William_Shakespeare
https://it.wikipedia.org/wiki/Teatro_elisabettiano
https://it.wikipedia.org/wiki/Come_vi_piace
https://it.wikipedia.org/wiki/Attribuzione_delle_opere_
di_Shakespeare
https://it.wikipedia.org/wiki/Re_Lear
http://disf.org/giovanni-paolo-ii-creazione-esistenza
https://it.wikipedia.org/wiki/Uovo_cosmico
https://it.wikipedia.org/wiki/Pelagianesimo
https://it.wikipedia.org/wiki/Orfismo
https://en.wikipedia.org/wiki/Bhagavad_Gita
https://it.wikipedia.org/wiki/Mircea_Eliade

https://it.wikipedia.org/wiki/Erebo

https://it.wikipedia.org/wiki/Eros

https://it.wikipedia.org/wiki/Uovo_cosmico

http://www.ilcerchiodellaluna.it/central_Simboli_uovo.h
tm

https://it.wikipedia.org/wiki/Mircea_Eliade
https://it.wikipedia.org/wiki/Big_Bang
https://it.wikipedia.org/wiki/Edwin_Hubble
https://it.wikipedia.org/wiki/Erwin_Schrödinger
https://it.wikipedia.org/wiki/Vedānta

31 ABOUT THE AUTHOR

A journalist admitted to Italian National Order since 1987, Carolina Paris (pseudonym of Carolina Monica Cirillo) works as a freelance with the most important Italian publishing groups.

In almost thirty years of activity she has authored hundreds of articles concerning technology, lifestyle, news, fashion, economy.

Passionate as she is about archeology, biblical studies, numerology, astronomy, genetics, Monica has a secret wish: to remain closed (alone) in the great Khufu's Pyramid for at least a day and a night.